AF589375

Crop Improvement in Vegetables

The Author

Dr. Navnath G. Kashid received his master degree in Botany from Dr. Babasaheb Ambedkar Marathwada University Aurangabad, Maharashtra in 1997 and Ph.D. from same university in 2004. He was pursuing doctoral work at Department of Botany Dr. Babasaheb Ambedkar Marathwada University Aurangabad. He is working as an Assistant Professor in Botany, Vasant Arts, Commerce and Science College Kaij, Dist. Beed, Maharashtra since June 1999. He is actively engaged in research and teaching under graduate students over last 17 years. His research interests are the areas of crop mutation breeding, molecular biology. He has more than sixteen years of diverse leadership experience as an academician and researcher with significant achievements in the field of plant genetics, mutation breeding for high yield and disease resistance crop plants.

He has authorized 45 research papers in national and international journals. He has guided two students for M.Phil degree and 04 students are doing Ph.D. He has associately published book "Problems in Genetics". with Rut printers' publication Jalna (Maharashtra). He is a life member of Indian science congress Association, Indian Botanical Society India; Marathwada Botanical Society Aurangabad (Maharashtra) and Indian society of genetics and plant breeding.

Crop Improvement in Vegetables

Dr. Navnath G. Kashid

2017

Scholars World

A Division of

Astral International Pvt. Ltd.

New Delhi – 110 002

Cataloging in Publication Data--DK
Courtesy: D.K. Agencies (P) Ltd. <docinfo@dkagencies.com>

Kashid, Navnath G., author.
Crop improvement in vegetables / Dr. Navnath G. Kashid.
pages cm
Includes bibliographical references.
ISBN 978-93-86071-44-6 (International Edition)

1. Okra--Mutation breeding--India. 2. Crop improvement--India. I. Title.

SB351.O5K37 2017 DDC 635.6480954 23

Published by : **Scholars World**
A Division of
Astral International Pvt. Ltd.
– ISO 9001:2015 Certi ied Company –
4736/23, Ansari Road, Darya Ganj
New Delhi-110 002
Ph. 011-4354 9197, 2327 8134
E-mail: info@astralint.com
Website: www.astralint.com

Dedicated to my beloved Parents

Mrs. Putalabai Gangadhar Kashid
and
Mr. Gangadhar Bhanudas Kashid

Acknowledgements

The book **Mutagenesis for crop improvement in vegetable plant** include several agronomic practices; biochemical experiments which make researcher and students easier in the study on induced mutation breeding is the main purpose of this book.

I wish to express my deep sense of gratitude to those individuals who helped me directly or indirectly during the completion of this work. Especially I am indebted to Prof. Dr. Vijay Kothekar Rtd. Head Dept of Botany, Dr Babasaheb Ambedkar Marathwada University Aurangabad Maharashtra, who allowed me to do this work. My sincerest thanks are also due to Prof. Dr. D S Mukadam, Prof Dr. Arvind Dhabe, Dr. Millind Sardesai, Dr Narayan Pandhure, Dr. Millind Jadhav, Dr. Mukund Kulthe, Dr. Subhash More, Dr. Santosh Talekar who suggest several things of interest time to time

I also wish my appreciation Dr. Madhav Fawde, Principal, Vasant Arts, Comerce and Science College Kaij, Dist.Beed, Maharashtra. I am really thanks to my all collegue who helps me several things in this work. I wish highest appreciation to my wife Suvarna, daughter Aditee and son Viren who support and provide their valuable time to make this work.

I wish to thanks everyone at the Daya publishing house, A division of Astral International Pvt.Ltd, New Delhi who worked on this book. I am grateful to Classic computer services New Delhi for typing and Thomson India Limited for printing this book.

Dr. Navnath G. Kashid

Preface

Mutation breeding is the purposeful application of mutations in plant breeding area. The knowledge of the extent to which the desirable characters with economic values are heritable is a prerequisite for any crop improvement programme. Breeders have continually retained their interest in the grouping to the germplasm and the pedigree of the selected cultivars since the information might be particularly helpful in effective breeding strategy determination. For this purpose, inducible mutation using chemical or physical mutagens, is a suitable source of producing variation through mutation breeding procedure which can produce several improved mutant varieties with high demanding economic value.

Mutagenesis techniques even the newer ones have resulted in many impressive mutant varieties. The number of such varieties is high across plant species. This book provide details of mutagenesis for crop improvement in vegetables. Emphasis is given on induced mutation breeding, morphological variation, agronomic characters and biochemical investigation of viable mutants in vegetable crops. This book gives an over all glimps of how mutagenesis techniques can be successful for the improvement of vegetable crops. It is hoped that this book will be of great value to students, teachers and researchers in vegetable science, plant breeding as well as reader interested in plant breeding and crop improvement.

The mutation breeding has become an alternative to conventional breeding since last three decades with sole objective of developing better cultivars of economically important crops.

The conventional plant breeder depends solely on the spontaneous mutations. However the frequency of spontaneous mutation is too low to become the factors contributing to them. An increase in mutation rate by mutation agent increases the chance occurrence of desired mutation and help in accelerating the breeding process as compared to conventional of developing novelties.

Dr. Navnath G. Kashid

Contents

List of Figures and Graphs

List of Tables

Chapter 1

General Introduction of Plant Mutation Breeding

The vegetables are an important component of global agriculture. It is a good source of energy, fats, vitamins, and fibers in human diet. The major cultivated vegetable species providing vitamins, fats and fibers comprise tomato, potato, chilli, brinjal and cauliflower. Among these the okra and tomato provide the high fiber content in human diet.

Okra or 'Lady's Finger', commonly known as "bhindi" in India, comprises one of the most important fruit vegetables grown throughout the tropics and warmer parts of the temperate zone. It is widely cultivated as a summer season crop in North India and also as a winter crop in Gujarat, Andhra Pradesh, Karnataka and Tamil Nadu. Okra is a hot weather tropical, low land crop, susceptible to drought and low night temperature. Thus it grows well in areas where day temperature remains between 25°C to 40°C and that of night over 22°C. It has been reported to have an average nutritive value (ANV) of 3.21 higher than tomato, egg plant and most cucurbits except bitter gourd (Grubben 1977).

Okra botanically described as *Abelmoschus esculentus* L. Moench. is a member of family Malvaceae. The genus *Abelmoschus* in fact was established by Medikus in 1787. However, most authors followed de Candolle (1824) and treated it as a section of *Hibiscus*. It was Hochreutiner (1924) who reinstated the genus *Abelmoschus* of Medikus, stating that calyx, corolla and stamens are fused together at the base and fall as one piece after anthesis.

Though the genus is of Asiatic origin, the origin of cultigen *Abelmoschus esculentus* is variable. The different centers of origin for okra comprise India (Masters 1875), Ethiopia (Candolle 1883), West Africa (Chevalier, 1940) and tropical Asia

(Grubben 1977), Howerver, Zeven and Zhukovasky (1975) believed it to have originated from India. This view gets its strength from the Sanskrit words, Tindisha and Gandhmula found to designate bhindi.

Records of okra production dates back to 12th century of Egypt, where it was grown widely. It was introduced to USA via French, Brazilian and African slaves. In India it moved from North Africa. Some of the okra growing countries of the world include Africa, Brazil, India, Pakistan, Bangladesh, United states of America and France. It is grown quite commonly in many other European and Asian countries. In India okra is commonly grown in states of Gujarat, Maharashtra, Andhra Pradesh, Uttar Pradesh, Tamil Nadu, Karnataka, Haryana and Punjab.

Table 1a: Distribution of Abelmoschus Species in different Phytogeographical Regions of India

Sl.No.	*Species*	*Distribution*
1.	*Abelmoschus angulosus*	Tamil Nadu
2.	*Abelmoschus cancellatus*	Uttaranchal, Himachal Pradesh, Utter Pradesh, Orissa.
3.	*Abelmoschus crinitur*s	Uttaranchal, Madhya Pradesh, Orissa.
4.	*Abelmoschus ficulneus*	Jammu and Kashmir, Rajasthan, Madhya Pradesh, Chhattisgarh, Maharashtra, Tamil Nadu, Andhra Pradesh, Uttar Pradesh.
5.	*Abelmoschus manihot* var. *tetraphyllus*	Rajasthan, Madhya Pradesh, Chhattisgarh, Maharashtra, Uttar Pradesh. Orissa.
6.	*Abelmoschus manihot,* var. *pungens*	Uttaranchal, Himachal Pradesh, Jammu and Kashmir, Assam, Andman and Nicobar Islands.
7.	*Abelmoschus moschatus ssp. moschatus*	Uttaranchal, Orissa, Kerala, Karnatka, Andman and Nicobar Islands.
8.	*Abelmoschus moschatus ssp. tuberosus*	Kerala and parts of western ghats in Tamil Nadu
9.	*Abelmoschus tuberculatus*	Rajasthan, Madhya Pradesh, Maharashtra, Uttar Pradesh.

Source: Bisht and Bhat, 2006.

Bhindi has a vast potential as one of the foreign exchange earning crops and accounts for about 60 per cent of the export of fresh vegetables, excluding potato, onion and garlic. The important countries like middle east, western Europe and USA, import bhindi from others (Sharma and Arora 1993). Such popularity on the part of okra could become possible mainly due to the presence of few positive attributes in that system. Such features include:

1. Short duration of the crop
2. Photoinsensitive nature
3. Low fibre content and
4. The ability of the plant to grow during any period of year.

Table 1b: State-wise Area, Production and Productivity of Vegetables

State/UT's	Area (in '000ha)					Production (in '000mt)					Productivity (in mt/ha)				
	1991-92	2001-02	2010-11	2011-12	2012-13	1991-92	2001-02	2010-11	2011-12	2012-13	1991-92	2001-02	2010-11	2011-12	2012-13
West Bengal	456.0	1139.0	1349.7	1330.94	1347.96	4680.0	18075.3	26725.5	23415.69	25466.81	10.3	15.9	19.8	17.6	18.9
Uttar Pradesh	576.7	777.9	829.4	852.09	912.66	9627.3	15044.8	17679.4	18563.75	19571.56	16.7	19.3	21.3	21.8	21.4
Bihar	843.3	578.9	845.0	857.0	861.8	8643.1	8022.9	14630.2	15552.4	16325.7	10.2	13.9	17.3	18.1	18.9
Madhya Pradesh	176.4	136.4	283.7	507.0	612.8	2221.0	1817.5	3698.6	10084.0	12574.0	12.6	13.3	13.0	19.9	20.5
Andhra Pradesh	155.2	222.5	651.2	661.0	686.1	1452.6	2586.7	11847.6	12025.3	12104.7	9.4	11.6	18.2	18.2	17.6
Gujarat	114.6	232.2	515.9	517.6	537.6	1667.9	3278.2	9379.5	10049.8	10520.7	14.6	14.1	18.2	19.4	19.6
Odisha	710.3	643.4	553.8	690.1	688.1	7275.0	7447.4	7790.1	9520.6	9464.0	10.2	11.6	14.1	13.8	13.8
Maharashtra	241.1	402.4	611.0	591.0	474.0	4171.3	5128.3	7504.0	8778.0	8008.0	17.3	12.7	12.3	14.9	16.9
Tamil Nadu	889.3	213.8	277.3	306.7	277.8	3796.9	5444.6	8279.9	9068.5	7897.9	4.3	25.5	29.9	29.6	28.4
Karnataka	351.1	358.1	466.3	454.7	436.6	3673.2	4173.2	9056.4	7662.5	7841.9	10.5	11.7	19.4	16.9	18.0
Haryana	60.8	150.4	346.4	356.8	360.3	877.0	2151.9	4649.3	5068.4	5011.3	14.4	14.3	13.4	14.2	13.9
Chhattisgarh		104.1	345.8	351.6	380.7		1355.3	4248.8	4582.6	4993.9		13.0	12.3	13.0	13.1
Jharkhand		158.5	259.5	261.2	321.5		1736.3	4112.4	3902.6	4325.4		11.0	15.8	14.9	13.5
Punjab	84.5	135.0	174.1	178.2	184.1	1450.0	2275.6	3585.8	3674.5	3782.6	17.2	16.9	20.6	20.6	20.5
Kerala	202.1	114.3	149.5	149.1	146.1	3229.1	2541.9	3392.7	3626.0	3446.9	16.0	22.2	22.7	24.3	23.6
Assam	222.4	237.4	260.1	266.0	278.7	2132.3	2935.2	2925.5	3045.6	3415.1	9.6	12.4	11.2	11.4	12.3
Himachal Pradesh	38.7	34.6	80.4	85.7	79.5	476.0	639.1	1474.9	1561.5	1521.1	12.3	18.5	18.3	18.2	19.1
J&K	180.3	50.8	69.7	63.1	63.1	745.0	728.9	1559.1	1395.5	1395.5	4.1	14.4	22.4	22.1	22.1
Uttarakhand	57.1	93.8	85.8	89.29	88.03	617.6	737.3	1030.9	1066.71	1059.57	10.8	7.9	12.0	11.9	12.0
Rajasthan	62.9	99.3	140.3	181.7	224.4	307.0	432.5	885.0	1287.4	873.5	4.9	4.4	6.3	7.1	3.9
Tripura	30.3	31.3	36.0	34.2	45.1	306.8	353.2	532.3	552.6	754.1	10.1	11.3	14.8	16.2	16.7

Contd...

Table 1b–*Contd...*

State/UT's	*Area (in '000ha)*					*Production (in '000mt)*					*Productivity (in mt/ha)*				
	1991-92	*2001-02*	*2010-11*	*2011-12*	*2012-13*	*1991-92*	*2001-02*	*2010-11*	*2011-12*	*2012-13*	*1991-92*	*2001-02*	*2010-11*	*2011-12*	*2012-13*
Delhi	55.0	111.0	29.8	27.9	27.9	627.8	747.4	496.8	466.7	439.3	11.4	6.7	16.7	16.7	15.7
Meghalaya	25.9	35.7	41.8	39.5	40.5	219.2	265.9	356.5	385.0	403.4	8.5	7.4	8.5	9.8	10.0
Mizoram	6.8	17.5	37.4	39.3	31.8	31.8	44.1	115.6	221.1	236.7	5.3	6.5	6.6	5.9	6.0
Manipur	11.8	10.6	22.2	20.8	21.7	50.3	66.1	236.5	200.3	219.8	4.3	6.2	10.7	9.6	10.1
Nagaland	8.2	26.3	10.7	33.0	26.0	66.9	286.0	79.4	222.6	207.7	8.2	10.9	7.4	6.7	8.0
Sikkim	7.6	14.2	23.9	25.0	25.6	46.1	60.0	120.9	127.7	132.5	6.1	4.2	5.1	5.1	5.2
Goa		7.6	5.7	6.5	6.7		76.0	57.8	78.2	80.5		10.0	10.1	12.0	12.1
Andaman & Nicobar	3.4	3.1	5.7	6.3	6.4	13.2	15.8	34.5	43.2	44.2	3.9	5.1	6.1	6.8	6.9
Arunachal Pradesh	17.1	20.8	4.2	6.3	1.5	79.9	83.9	38.5	83.5	37.6	4.7	4.0	9.2	13.2	24.7
Puducherry	2.3	3.7	0.6	0.6	1.5	22.3	54.2	8.8	7.5	25.0	9.7	14.6	14.7	13.6	17.1
Dadra & Nagar Haveli	1.5	1.5	1.1	1.1	1.1	13.6	13.5	5.5	5.5	5.5	9.1	9.0	5.0	5.0	5.0
Lakshadweep	0.4	0.2	0.4	0.3	0.3	0.4	0.2	14.1	0.3	0.3	1.0	1.0	35.3	1.2	1.3
Daman and Diu	0.1	0.1	0.0	0.0	0.0	0.3	1.1	0.0	0.0	0.0	3.0	11.0			0.0
Chandigarh	0.3	0.1	0.1		0.0	11.1	1.7	1.7		0.0	37.0	17.0	17.0		0.0
TOTAL	5136.7	5016.7	8494.6	8989.6	9205.2	53852.0	70546.7	146554.5	156325.5	162186.6	10.5	14.1	17.3	17.4	17.6

Source: Director of Horticulture/Agriculture of respective State/UT's.

Table 1c: State-wise Area, Production and Productivity of Okra

State	2010-11			2011-12			2012-13		
	Area '000 ha	Production '000 mt	Productivity mt/ha	Area '000 ha	Production '000 mt	Productivity mt/ha	Area '000 ha	Production '000 mt	Productivity mt/ha
Andhra Pradesh	78.9	1184.2	15.0	80.1	1202.0	15.0	74.25	1113.81	15.0
West Bengal	74.0	862.1	11.6	74.4	863.5	11.6	74.60	869.10	11.7
Bihar	58.5	788.3	13.5	59.0	825.3	14.0	59.24	854.22	14.4
Gujarat	54.5	592.5	10.9	65.4	717.3	11.0	65.66	723.33	11.0
Odisha	70.0	651.8	9.3	67.6	651.8	9.6	67.04	593.93	8.9
Jharkhand	30.0	421.7	14.1	31.8	444.7	14.0	32.52	447.40	13.8
Maharasthra	19.0	224.0	11.8	21.0	317.0	15.1	22.00	328.00	14.9
Madhya Pradesh	10.1	60.3	6.0	23.6	310.0	13.1	25.74	297.00	11.5
Chhatisgarh	25.2	249.1	9.9	25.8	252.2	9.8	26.47	269.18	10.2
Assam	11.1	147.8	13.4	11.3	154.4	13.7	11.52	169.11	14.7
Uttar Pradesh	11.1	128.8	11.6	11.5	136.0	11.8	12.44	159.30	12.8
Haryana	18.2	145.3	8.0	18.8	139.7	7.4	18.03	154.09	8.5
Others	37.7	328.2	8.7	28.2	245.4	8.7	41.3	371.8	9.0

Source: Director of Horticulture/Agriculture of respective State/UT's.

Table 1d: All India Area, Production and Productivity of Vegetables

Year	*Area (in '000 ha)*	*Production (in '000 mt)*	*Productivity (in mt/ha)*
1991-92	5593	58532	10.5
2001-02	6156	88622	14.4
2002-03	6092	84815	13.9
2003-04	6082	88334	14.5
2004-05	6744	101246	15.0
2005-06	7213	111399	15.4
2006-07	7581	114993	15.2
2007-08	7848	128449	16.4
2008-09	7981	129077	16.2
2009-10	7985	133738	16.7
2010-11	8495	146555	17.3
2011-12	8989	156325	17.4
2012-13	9205	162187	17.6

Source: Director of Horticulture/Agriculture of respective State/UT's.

Table 2a: All India Area, Production and Productivity of Okra

Year	*Area (in '000 ha)*	*Per cent of Total Veg. Area*	*Production (in '000 mt)*	*Per cent of Total Veg. Production*	*Productivity (in mt/ha)*
1991-92	222.0	4.0	1886.5	3.2	8.5
2001-02	347.2	5.6	3324.7	3.5	9.6
2002-03	329.2	5.4	3244.5	3.8	9.9
2003-04	353.1	5.6	3631.4	3.9	10.3
2004-05	357.3	5.3	3512.4	3.5	9.8
2005-06	391.8	5.5	3974.6	3.6	10.1
2006-07	396.0	5.2	4070.0	3.5	10.3
2007-08	407.0	5.2	4179.0	3.3	10.3
2008-09	432.0	5.4	4528.0	3.5	10.5
2009-10	452.5	5.7	4803.3	3.6	10.6
2010-11	498.0	5.9	5784.0	3.9	11.6
2011-12	518.4	5.8	6259.2	4.0	12.1
2012-13	530.8	5.8	6350.3	3.9	12.0

Source: Director of Horticulture/Agriculture of respective State/UT's.

Table 2b: Area and Production of different Vegetable Crops in India

Crops	*Area (ha)*	*Production (tonnes)*	*Productivity (t/ha)*
Ashgourd	2497	15326	6.13
Beet root	2164	36260	16.75
Bittergourd	26004	162196	6.23
Bottle gourd	116939	1428296	12.21
Brinjal	299770	3124487	10.46
Cabbage	113450	1631690	14.38
Caspium	4783	42230	8.83
Carrot	20124	2870007	14.26
Cauliflower	238632	3394897	14.22
Chilli	441050	919339	12.02
Cowpeas	23012	133587	5.80
Cucumber	16288	105690	6.40
Dolichos	600	9000	15.00
French bean	4268	24778	5.80
Garlic	41842	218985	5.23
Leaf vegetables	111840	731158	6.53
Longmelon	500	7500	15.00
Luffa	73273	685224	9.35
Methi	13510	83600	6.19
Muskmelon	28484	625414	21.95
Okra	399684	2326616	6.28
Onion	280915	3181067	11.32
Other bean	688645	349490	0.52
Other cole crops	16007	195424	12.20
Other gourds	109846	659239	14.32
Peas	146991	2105686	12.35
Pumpkin	43317	532779	11.91
Radish	67345	802529	12.77
Ridge gourd	10040	128310	15.85
Tomato	290279	4603446	15.85
Watermelon	16194	205884	12.71

Modern okra cultivars are day neutral but some can display short day response as well. The temperature and photoperiod influence the okra flowering. It requires optimum temperature of 24°C to 27 °C. (Nath *et al.*, 1987). It is a good candidate for rotation with winter crops and early spring crops.

Okra is generally grown on soil from sandy loam to clay loams. Sandy loams are considered best for early crop in spring as where clay loam is best for good yield. Optimum pH is between 6.0 to 6.8 but okra can be grown successfully on saline soils upto 6mmhos/cm EC (Malik *et al.*, 1981).

The most important centers which are actively engaged in the improvement of okra in India are: Department of vegetable crops, Punjab Agriculture University, Ludhiana; Department of Botany, Marathwada Agriculture University, Parbhani, Maharashtra; Division of vegetable crops, IIHR, Bangalore, Karnataka and Department of Horticulture, Tamil Nadu Agriculture University, Combatore, Division of plant introduction, IARI, New Delhi, Department of vegetable crops, HAU, Hissar etc. However all the agriculture universities and vegetable research institutes in our country do carry some work on okra to meet their local needs.

The Economic Importance of Okra

Okra is a vegetable crop cultivated throughout India. Its immature fruits are generally cooked as vegetable. The extract of okra is used in manufacturing jaggery. The okra plant fibre is used by paper industries (Nadkarni 1927).

The okra fruits having long pod with ribbes and spineless condition are specially valued for tender and delicious vegetable preparation carrying immense medicinal value. Okra soups and stews are popular dishes; seeds are roasted and used as a substitute for coffee. (Martin 1982). Due to high iodine content the okra fruits are considered useful for control of goitre.

Okra fruits are considered better for people suffering from renal, colic, leucorrhoea and general weakness. Bland mucilage is also used for control of dysentry and as a clarifying agent in preparation of cur.

Okra leaves are large, broad, heart shaped with lobes. They are used in preparation of medicines to reduce inflammation in Turky (Mehta 1959).

Okra is a good source of carbohydrates, minerals, vitamin-A and vitamin-C. It is stuffed with meat or cooked into soup. It is also canned and dehydrated for off-season consumption by army at high altitudes and for exports (Table 3).

Table 3: Nutritional Composition-Amount 100g Edible Portion of Okra

Calories	35.0	Moisture(g)	89.6
Carbohydrates (g)	6.4	Protein (g)	1.9
Fat(g)	0.2	ibre (g)	1.2
Minerals g()	0.7	Phosphorus (mg)	56.0
Sodium (mg)	6.9	Sulphur (mg)	30.0
Calcium (mg)	66.0	Iron (mg)	0.35
Potassium (mg)	103.0	Magnesium (mg)	53.0
Copper (mg)	0.19	Riboflavin (mg)	0.01
Thiamine (mg)	0.07	Nictonic acid (mg)	0.06
Vitamin C (mg)	13.10	Oxalic acid (mg)	8.0

Source: Gopalan *et al.*, 2007.

The Relevance of Mutation Breeding

The mutation breeding has become an alternative to conventional breeding since last three decades with the sole objective of developing better cultivars of economically important crops. The conventional plant breeder depends solely on the spontaneous mutations. However the frequency of spontaneous mutation is too low to become the basis for systematic breeding, even taking into account the internal and external factores contributing to them. An increase in mutation rate by mutagenic agent increases the chance occurrence of desired mutations and helps in accelerating the breeding process as compared to conventional method of developing novelties.

Ionizing radiations and various chemical mutagens provide handy tools to enhance natural mutation rate, thereby enlarging the genetic variability and increasing scope for obtaining desired mutants. Ionizing radiations such as X-rays, gamma rays, neutrons and some proven chemical compounds have been found to be quite effective as mutagenic agents. Mutational studies relating to alterations of mutation frequency and spectrum are expected to achieve a certain measure of directed mutagenesis (Swaminathan and Sharma 1969).

Mutations according to Hugo de Vries (1901) represent heritable variations and have an important role in evolution. The scientific study of mutations started in 1910 when Morgan started his work on *Drosophila* and reported white eyed males among red eyed male population. The mutagenic effect of X-rays was demonstrated in *Drosophila* for the first time by Muller (1927) and subsequently in barley and maize by Stadler (1928). Besides radiations the mutagenic potential of some chemicals was also decisively demonstrated (Auerbach 1947). It was realized that there were several chemicals which could be used for inducing mutations viz Mustard gas, Ethyl urethane, formaldehyde, Alkly sulphonates, Nitrosoamines, Aziridines and Epoxides.

Several hundreds of crop varieties have been developed through induced mutagenesis and released till to date. In addition to these several varieties were developed in ornamental plants through similar methods. A remarkable change can be seen after going through the practical results monitored by the FAO/IAEA. The number of varieties in crops developed through the use of induced mutation increased from 98 to 335 and in ornamentals increased from 47 to 271 during 1973 to 1984 (IAEA mutation breeding Review 1985) which suggests and demonstrates the enormous potential of mutation breeding in crop improvement.

Induction of mutations is an important method for improving the specific characteristic and rectifying one or few undesirable characters in the existing commercial varieties. The created variability can further be utilized in selection or hybridization to evolve new cultivars.

The details of the prospects and utilization of induced mutations have been compiled and published as review papers by Gustafsson (1938, 1940, 1947, 1969), Konzak (1956), Mackey (1956), Sparrow, Nybom (1961), Gaul (1958, 1964 and 1965), Swaminathan (1963) and Gottschalk and Wolff (1983). The most important publications in this field regarding the list of released, commercialized and approved mutant varieties in different crop plants are by Sigurbjornsson and Micke (1969) and Broertjes and Van Harten (1978). In addition to crop improvement, the genetic variability induced by physical and chemical mutagens also help in analyzing and understanding genes, their regulation and organization (New combe 1971).

Chopra and Sharma (1985) have reported beneficial prospects of mutation breeding in India and provided a long list of improved varieties isolated through mutagenesis. Datta (1994) has indicated the utility of mutation technique in producing large number of new promising varieties in different ornamentals like chrysanthemums and roses.

The genetic variability in mutagen treated population could be obtained through the alterations in genetic material. Such alterations in genetic material may be of a micro or macro type. The macrogenetic changes mostly manifest themselves in the form of chromosomal abnormalities, *viz.* sticky chromosome, multivalent associations, micronuclei, bridges etc. in the pollen mother cells. These chromosomal abnormalities may result in development of new linkage groups which may induce genetic variation in plants.

Mutations are grouped into two major categories on the basis of their phenotypic manifestations:

1. Micromutations – these involve changes in quantitative traits and can be measured at the level of population using various statistical parameters, such as, character mean, variance, heritability etc. and,
2. Macromutations – with large changes in the characters which can be detected even without instrumental help at the level of individual plant.

The interest in micromutations for generating polygenic variability increased after Brock (1965) proposed the hypothesis of induction of quantitative variability through mutagenic treatment. Micromutations produce genetic variability in quantitative traits of the crop plants. Hence, they deserve full attention of plant breeders. Such mutations should be useful for improving quantitatively inherited traits (such as yield) without disturbing the major part of the genotype and the phenotypic architecture of the crop.

In species with generative propagation, preferably the seeds are subjected to mutagenic treatment. The mutation rate increases with increasing dose, simultaneously the non-genetic primary damage also increases resulting in a decreasing ability to germinate, increasing inhibition of growth and other morphological and physiological disorders during plant growth with increasing dosage.The proportion of chromosome mutations with conspicuous morphological and physiological changes usually increase. Such study have led to the development of a number of basic concepts, such as the concept of LD-50, dose rate and dose fractionation effects in relation to physical and chemical mutagens, the transitory state of premutational lesion finally getting fixed into mutational changes and existence of sieves like diplontic and haplontic selections (Gual 1961).

The induced mutations display a pleiotropic effect due to which several features are changed only one of which however is desired by the breeder. The cross of two mutants can occasionally break undesired pleiotropic effect and valuable transgressions in other features can be achieved. Since majority of point mutations are recessive, the search for deviating phenotypes does not begin until the M_2 generation. The determination as to whether a detected variant is a mutant or non-heritable modification can only take place in M_2 generation after testing its progeny. The significance of the differences in quantitative features can be ascertained by applying biometrical method.

Induced mutations offer fewer prospects for improvement of cross fertilizing species than for self fertilizing species. This is partly because of difficulty of selecting, incorporation and maintaining recessive mutations in such a population and because the plant breeding problems in cross-fertilizing species are more often the problem of handling the existing variability than lack of variability perse. However several successful plant improvement ventures have been undertaken with cross-fertilizing species exposed to mutagenic agents (Sigurbjornsson and Micke 1969) and the induction of mutation has been accepted as a useful tool in plant breeding programmes. In case of vegetable crop the utility potential of mutation breeding has been nicely demonstrated by different researchers for increasing the productivity in different systems like Tomatoes (Wokes and Organ, 1943), Brinjal (Datar and Astapure 1984), Chilli (Gupta and Yadav 1984) and in Custer bean (Basha and Rao, 1988).

It is very much believed by the mutation breeders that the desirable mutants in different vegetables would be able to contribute effectively towards the fibre and protein production besides getting the induced genetic variability for the much sought after disease/pest/insect resistance. Keeping this end in view it was thought very much logical to initiate the work in regard to the mutation breeding of okra under our local conditions for broadening the genetic base of the crop and for evolving useful genes having direct bearing on the breeding programme of the okra system.

Chapter 2

Literature Study of Mutation Breeding in Vegetable Plants

The present review intends to cover earlier published cytogenetical and induced mutational information on okra (*Abelmoschus esculentus* L. Moench) besides the related aspects.

Taxonomy

Okra or Lady's finger, commonly known as bhindi, belongs to genus *Abelmoschus* which was established by Medikus in 1787. de Candole (1824) treated it as a section of *Hibiscus* and assigned it to family Malvaceae.

The genus *Abelmoschus* is of Asiatic, variable India (Master 1875), Ethiopia (Candole 1883) and West Africa (Chevalier 1940) origin.

There are 30 species under the genus *Abelmoschus* in the world and four species in the New world (Joshi *et al.* 1975).

Centre of Origin and Spread of Okra

The okra (*Abelmoschus esculentus* L. Moench) has originated in Abyssinia or the presently Ethiopia and the upper Nile river region of Sudan. It has been recorded from 12th century in wild form.

It shows a long history of cultivation. In India probably it arrived at the end of 19th century. The okra cultivation is observed in a number of countries like Brazil, U.S.A., France, Africa, Bangladesh and India.

Plant Description

The cultivated okra plant having short-branched habit comprises a warm season dicot plant. It is cultivated throughout the year.

The okra plant is sensitive to cold conditions. It belongs to same family as those of cotton and hibiscus. The quantitative characteristics of okra are valuable and are used in vegetable and medicine preparations.

Stem

The stem of okra is unbranched. Although branched cultivars are observed in commercial field. Okra stem is semi woody and sometimes pigmented with green and reddish tings colour with spines. The dimensions and development of stem including branching are influenced strongly by the environment.

Stem length of commercial cultivars varies from 95 to 195 cm and that of dwarf cultivars is less than one meter. The diameter varies from 2 to 10 cm. The stem length is determined by number and length of internodes. Both tall and short plants with many internodes possess thick stem because of positive association between internodal number and stem thickness. The stem is medium tall with small internodes with splashes and pigmentation (Thakur and Arora 1985).

Internode

The length of the stem in okra is proportional to the number of nodes on the stem (Yashvir 1975). The internode elongates to 7.8 to 20.20cm length (Anonymous-1980). Elongation of the internodes takes place rapidly at the base, which later gets shifted progressively up the internode and ceases at the base when internode reaches 65 per cent of its final length

Branching

Okra generally possesses small branches on main stem. Variants as regards branching in cultivated and wild type okra are known. There is a correlation between number of branches and yield in okra (Ramana Rao 1991).

Length of branches varies from few centimeters to a length longer than the main stem. Branches are concentrated near the basal portion or over the entire plant.

Root

Tap root system in okra maintains water balance between plant and soil. The roots grow 150 to 225 cm deeper in soil. Root develops multiple branches (Krishnaswami 1967).

During early plant development numerous strong lateral roots originate from enlarged tap root. The laterals spread widely from 60 to 120 cm in soil.

Leaves

Seedlings emerge from soil and cotyledons unfold and the first pair of leaves becomes visible at the tip of the shoot axis. Leaves are produced in alternate phyllotaxy. Leaves are broad with five lobes or are reduced into 3 to 5 lobes (Basha and Rao, 1988).

Leaves are green, small, deeply lobed having sparse hairs and vary in intensity of green colour. Cultivars having light green, dark green leaves are on record. The angle of petiole on the main stem could be semierect, horizontal or having down. Petiole is sparsely haired (Thakur and Arora 1985).

In commercial cultivars of okra the leaves are simple with shallow incision having 16.2 cm length and 15.3 cm width (Singh et.al, 1974).

Leaves are deeply lobed with narrow leaflets in the top 1/3 portion of the plant.

Inflorescence

Okra is having an axillary solitary inflorescence and the flowers are developed at the axil of leaf having similarity with that of hibiscus and cotton.

Flower

Okra flowers are large and showy with five petals (Singh 1973). The flowers are perfect and self fertile. After 6-8 nodes the flowers open once for a short period during day time.

Flowers are with monoadelphous stamens and make the emasculation and pollination easier. It is having self and cross pollination pattern (Randhawa and Purewal 1947). The natural cross pollination varies from 4.0 to 18.75 per cent, the average being 8.75 per cent. The okra flowers are bisexual with five sepals, five petals and pentacarpellary ovary.

Fruit (Pod)

Okra fruit is a long pod, comprised of several fused carpels with ribbes and a spineless feature. Colour of pod ranges from white, pale to dark green and even red coloured commercial cultivars also are known. Depending on cultivars, pod may be ribbed or rounded, smooth or spined, pointed or blunt, long or short. Number of pods per plant depends upon frequency of picking of pods (Grewal 1971). Rao and Ramu (1977) reported additive gene effects regarding number of pods per plant.

Dhillon and Sharma (1979) recorded increased number of pods through early flowering. Tunwar and Singh (1985) recorded okra pod with long slender feature besides, 5-ribbed with glossy smooth and scarlet red colour, which disappeared on cooking. Joshi *et al.* (1956) recorded okra pods with smooth dark green colour and with 18-20 cm length.Venkataramani (1952) recorded increased pod yield ranging from 5.4 to 14.5 per cent over better parent.

Seeds

Okra seeds are round, dark green to dark brown in colour with 5mm diameter. The seeds consist of seed coat and endospermous embryo.

The colour of okra seed coat judges the quality of seed in the yellow green and gray green seed types (Singh and Gill 1988). Hard seed coat of okra poses problem in germination (Manohar and Solanki 1969). The embryo in case of okra consists of enzymes, acids, alkalies, glucose, phosphate and the endogenous auxins.

Germplasm Collection

Germplasm collections have been made indigenously and the exotic sources are utilized in different parts of country.The major collection of okra is maintained by the NBPGR, New Delhi. IBPGR has assigned the NBPGR the global responsibility of base collection of okra.

Okra germplasm endemic to India and Bangladesh have been given special emphasis as regards collection and evaluation of okra. Large number of okra types including wild relatives have been collected and utilized. Germplasm collection is also attempted by Departments of vegetable crops of Agriculture Universities of Ludhiana, Hissar, Coimbtore, Parbhani, besides IIHR, Bangalore and IARI, New Delhi.

The earliest collection of okra line IC 1542, has been made from West Bengal. The germplasm collection of a large number of okra lines has been reported by Jha and Mishra (1955), Arumugam *et al.* (1975), Rao and Bidari (1976) and Mahajan and Sharma (1979).

It was noticed by Sandhu *et al.* (1974) that out of 94 lines screened, an accession EC 31830 'Asuntem koko' from Ghana could be the *Abelmoschus* Manihot L. Medicus spp. which was reported earlier by the Royal Botanical Gardens Kew. This accession in due course has contributed to the development of resistant CVS. Punjab Padmini and Punjab-7 at Ludhiana, Punjab.

Progress at Varietal Front

The commercial cultivation of okra started in India at the plant introduction section, Division of Botany, IARI, New Delhi after 1950. Okra cultivars were developed mainly on the basis of pod attributes, growing season, place of cultivation and development oriented characters.

Singh and Sikka (1953) introduced new cultivar "Pusa Makhmali" with high yielding potential.Singh (1957-58) introduced "Pusa sawani" from IARI, New Delhi, which was notified for general cultivation in whole of our country by central seed committee in 1969.

Tamil Nadu Agriculture University, Coimbtore in 1976 released new cultivar of okra 'CO1'with long pod with high yielding feature (Arumugam and Muthukrishanan 1977). In 1978, it again evolved "MDU-1" mutant variety with higher number of nodes on stem (Thandapani *et al.*, 1978).

Recognizing the value of hybrid Parbhani Kranti, YMV resistant variety was evolved for commercial cultivation by Jambhle and Nerkar in 1985. The hybrid base has been further windened by the release of more number of hybrids in last few years having better yield and appreciable disease resistance.

Cytology

The cultivated okra *Abelmoschus esculentus* L. Moench is polyploid in nature (Joshi and Hardas 1956). *Abelmoschus esculentus* (2n = 130) and *Abelmoschus turberculatus* (2n = 58) are on record.

Abelmoschus esculentus (2n = 130) evolved through hybridization between species with n = 39 and n = 36 by doubling of chromosome number. The genome

Table 4a: Chromosome Numbers (2n) Abelmoschus

Species	*Chromosome Numbers (2n)*	*Authors*	*Ploidy Level*
Abelmoschus esculentus	± 66	Ford 1938	
	72	Teshima 1933, ugale *et al.,* 1976 and kamalova 1977	
	108	Data and Naug 1968	2
	118	Krenke In: Tischer 1931	2
	120	Krenke In: Tischer 1931, purewal and Randhawa 1947	2
	122	Krenke In: Tischer 1931	2
	124	Kuwada 1957, 1966	2
	126-134	Chizaki 1934	2
	130	Skovsted 1935, Joshi and Hardas 1953, Gadwal In: Joshi and Hardas 1976, Gadwal *et al.,* 1968, Joshi *et al.,* 1974, singh and Bhatnagar 1975	
	131-143	Siemonsma 1982a, 1982b	2
	132	Medwedewa 1936, Roy and Jha 1958 1958	2
	± 132	Breslavertz *et al.,* 1934, Ford 1938	2
	144	Data and Naug 1968	2
Abelmoschus manihot	60	Teshima 1933 and Chizaki 1964	1
	66	Skovsted 1935, Kamalova 1977	1
	68	Kuwada 1957, 1974	1
Abelmoschus tetraphyllus	130	Ugale *et al.,* 1976	2
	138	Gadwall In: Joshi and Hardas 1976	2
Abelmoschus moschatus	72	Skovsted 1935, Gadwal *et al.,* 1968, Joshi *et al.,* 1974	1
Abelmoschus ficulneus	72	Kuwada 1966,1974, Joshi and Hardas 1954, Gadwal *et al.,* 1968, Joshi *et al.,* 1974	1
Abelmoschus tuburculatus	58	Kuwada 1966,1974, Joshi and Hardas 1953, Gadwal *et al.,* 1968, Joshi *et al.,* 1974	1
Abelmoschus angulosus	56	Ford 1938	1
Abelmoschus caillei	194	Singh and Bhatnagar 1975	3

Source: Report of an international workshop on okra genetic resources IBPGR, Rome, 5: 52-68. 1991.

of *Abelmoschus turberculatus* and *Abelmoschus esculentus* with n = 36 has been widely investigated in India and Japan.(Joshi and Hardas, 1956).

Ploidy

The perfect pairing of 36 chromosomes in okra at zygotene of meiosis-I has been observed by Ugale (1976). One of its genomes manifested a good homology with *Abelmoschus esculentus* and behaved like an amphidiploid.

Inspite of chromosomal homology between *Abelmoschus esculentus* and *Abelmoschus tetraphyllus*, the seed sterility has been distinctly recorded in the irhybrid (Dhapke and Meshram 1981). The chromosome pairing in the amphidiploid was normal with 129-134 per cell as against only 20-40 per cell in the F_1.

The induced amphidiploid showed regular meiosis forming 134 bivalents owing to preferential pairing of chromosomes. The species with highest chromosome number was used as female. Most of meiotic chromosomes remained as univalents in *Abelmoschus esculentus*

Genetics

Yield and Yield Components

In okra much genetic variability occurs for seed and pod yield. Malik (1986), recorded high heritability and genetic advance for pod length and pod diameter.

Rao and Ramu (1977) reported gene effects for number of pods, pod yield and seeds per pod. The dominant gene effects for days to anthesis, pod number, first flowering, number of pods per plant, plant yield have been reported by Sharma and Mahajan (1979).

Arora (1980) reported additive gene effects for number of ridges, seed weight, nodes per plant, days to marketable maturity, total yield and protein content.

Plant Growth Characteristics

Chlorophyll Deficiencies, Colour Difference

Singh *et al.* (1962) reported that seedling entirely devoid of chlorophyll could be controlled by single recessive gene and the colour and the margin of leaves is controlled by dominant gene.

Venkatramani (1952) inferred dark green colour of stem and the leaves to be dominant the light green while the green to be recessive to greenish red.

Shape Size and Number of Leaves

Arumugan and Muthukrishanan (1979) inferred the presence of additive and dominant gene effects for the number of branches and leaf number.

The leaf lobbing was found to be controlled by single complementary dominant gene by the same researchers.

Table 4b: Characteristics of Varieties and Hybrids

Variety	*Year of Realese*	*Plant Height (cm)*	*Days to Flowering*	*First Picking*	*Pod*	*Colour*	*Length of Pod (cm)*	*Yield a/ha*	*Characteristic Features*
Pusa Makhamali	Singh and Sikka (1995) IARI, New Delhi	180-220			Smooth 5-ridgid	Light green	15-50	80-100	Pod straight, Attractive, Tapered
Pusa Sawani	H. B. Singh (1957-58) IARI, New Delhi	120-180			Smooth hairy 5-ridgid	Dark green	18-20	125-175	Purple patch on petal grown whole country
CO-1	Tamil Nadu Agricultural University, Coimbtore (1978)	90-120			Glossy 5-ridgid	Scarlet red colour			Red colour fruit borer. Suitable for Tamil Nadu
MDU-1	Tamil Nadu Agricultural University, Coimbtore (1978)	75	33	43	Smooth 5-ridgid	Light green	20		Stem with pigementation, early flowering
Punjabi Padmini	Sharma (1982) Punjab Agri. University, Ludhiana	180-200	45-50	53-54	Smooth 5-ridgid	Dark green	15-20	100-125	Quick growing dark green fruit
Gujarath bhindi-1	Gujarat Agri. University, Ahmadabad (1983)	60	60	55-60	5-ridgid	Green	14-15	70	Broad leaves, with purple tings on veins suitable for Gujarat
Harbhajan	Dr. Saini *et al.* (1983-84), Agri. College Solan, Himachal Pradesh				Spineless 8-ridgid	Bright green			Long tapered fruit with 8-eadge suitable for Himachal Pradesh
Selection-2	H. B. Singh and *et al.* (1973-74) IARI, New Delhi	110	40-45	45-50	5-ridgid	Green		110	Purple patal narrow leaflet, Resistant to YVM

Contd...

Table 4b–*Contd...*

Variety	*Year of Realese*	*Plant Height (cm)*	*Days to Flowering*	*First Picking*	*Pod*	*Colour*	*Length of Pod (cm)*	*Yield a/ha*	*Characteristic Features*
Parbhani Kranti	Jamble and Nerkar (1985) Marathwada Universyti, Parbhani	110-120	40-45	55	Smooth tender 5-ridgid	Dark green	8-9	115	Red colour fruit borer. Suitable for Tamil Nadu
P-7	Tamil Nadu Thakur and Arora (1985) Punjab Agri. University, Ludhiana	105	40-45	54	5-ridgid	green		95	Pigemented petile with hair YVM-resistant
Set-10 Arka Anamika	Datt *et al.* (1984) IIHR, Banglore	100	50	55	Rough 5-ridgid	green		115	High yielding, leaves, stem, petiole are sparsely haired suitable for all India
EMS-8	Sharma and Arora (1989) Punjab Agri. University, Ludhiana	150			Sparsely hairy 5-ridgid	green		95	Tall plant, splashes of pigmentation on leaves. Suitable for Punjab.

Branching

Randhawa and Sharma (1988) reported single dominant gene and genetic advance for number of fruits and branches per plant.

Kulkarni and Thimmappaiah (1977) reported that depending upon parents involved, both additive and dominant gene effects are possible as regards plant height, fruit number and branches having marketable value.

Plant Height\

Although plant height has been regarded as a quantitative character, additive gene effects for days to flowering and plant height have been reported by Rao and Sathyavathi (1971).

Single additive gene affecting plant height, internodal length and branches have been reported by Singh (1979), Ramu (1976) and Sharma and Mahajan (1980).

Flowering and Maturity

Malik (1968) reported high heritability and genetic advance for days to flowering.

Rao (1977) reported additive and dominant effects for days to flowering. Some cultivars having short day response for flowering have also been noticed.

Kolhe and D'cruz (1966) observed mosaic pigmentation on parts like calyx and corolla.

Arora (1980) reported the presence of additive gene effect for days to flowering in new cultivars.

Pod Size and Shape

Pod size is influenced greatly by environmental effects. The number of ridges and pods per plant having marketable value are affected by single additive gene (Arora *et al.*, 1980).

Singh (1983) reported single gene for more fruit number, fruit weight and longer fruit with higher yield. The pod colour ranging from white through dark green and some cultivars with red colouration are also very much on record.

Protein and Vitamin Composition

Additive gene effects for sugar, vitamin-C and crude fibre content have been indicated by Malik (1968).

The presence of gene for seed weight, total yield and protein content is also reported (Arora 1980).

Breeding

Okra is a highly open pollinated type. The plants have perfect flowers and self fertility methods are used in okra breeding besides some current procedures which are directed towards developing cultivars and hybrids of okra.

Methods for Improving Cultivars

1. Selection
2. Hybridization

Selection

Selection generally refers to phenotypic selection of individual plant for the improvement of cultivars or population. For some specific trait, selected plants are harvested and their seeds are mixed in mass selection and they are utilized without mixing in pure line selection to form a new cultivar.

Mass selection for improvement of okra was used commonly at IARI New Delhi during early stages of improvement of the cultivar Pusa Makhamali.

Gujarat Bhindi was developed from pure line selection of an unknown bulk seed sample from IARI New Delhi.

Harbhajan, Sel-4, Arka Abhay are the new cultivars from the selection methods (Tanvar and Singh 1985).

Inter Specific Hybridization

Much success has been achieved in recent years in hybridizing different species of okra by using new breeding techniques. Successful interspecific hybridization in *Abelmoschus esculentus* (L.) Moench has helped generate considerable variability for isolating desirable genotypes (Joshi *et al.*, 1974).

A large number of interspecific crosses between the likely parental species have been attempted *A manihot* X *A. tburculatus* X *A. moschatus* (Gadwal *et al.*, 1968) have been attempted to help raise viable hybrids.

Mamidwar *et al.* (1971) studied interspecific cross in cultivated okra and three wild species namely *A. tetraphyllus, A. manihot* (L.) *A. manihot* (Medik). The causes of sterility in the hybrids were mostly chromosomal and genic differentiation.

The hybrid was highly sterile and produced fruits without seed (Jambhale and Nerkar 1982).

Interspecific cross transferring the gene for resistance to Mosaic fruit had been successfully carried out by using *A. manihot* parent (Thakur 1976 and Datta 1984).

Joshi and Gadwal (1968) reported the embryo and ovule culture in the inter specific crosses of okra.

Heterosis

Manifestation of heterosis for various economic traits has been reported in okra. Vijayraghawan and Warrior (1946) reported increase in fruit size, fruit length, fruit weight and number of fruits per plant in the F_1 hybrids.

Venkataramani (1952) reported increase in fruit yield ranging from 5.4 per cent to 14.5 per cent over the better parent in okra

Heterosis over mid parent has been reported by Singh (1983) for more fruit number, fruit weight and longer fruits besides higher marketable and total yield, tall plants, less number of nodes and days to first pod appearance.

Mutagenesis (General)

The main objective in okra breeding is to increase its pod yield, protein content along with other agronomic characters. This could be achieved through induced mutation which is an effective tool for creating new variability in plant breeding programmes.

In recent years lot of work has been done on the induction of mutations in different crop plants and this has resulted into significant crop improvement.

The experiments of Muller (1927) in *Drosophila* followed by Stadler (1928a, 1928b) in maize and barley established beyond doubt that progress in plant breeding can be expedited by using induced mutations for expanding the genetic variability of crop species. The work of Gager and Blackeslee (1927) and Goodspeed (1929) in *Datura stramonium* and *Nicotiana* confirmed the above findings decisively.

The mutagenic nature of chemicals was demonstrated for the first time by Auerbach (1943), Oehlkers (1943) and Rapoport (1946). Since the effects of chemicals were analogous to radiations, they were called as Radio-mimetic substances.

There is ample literature available in the filed of applied mutagenesis. The problems, methods and results of mutation breeding have been discussed at length by Gustafsson (1947 and 1969), Konzak (1956 and 1957), Sparrow (1961), Brock (1965), Swaminathan (1969 and 1971) and Gual *et al.* (1972).

Mutagens induce differential genetic and cytogenetic changes. Thus mutagenic effectiveness and efficiency will also depend upon the nature of mutations induced.

In case of sparsely ionizing radiations like gamma rays, the ratio of point mutations to chromosomal aberrations is much higher than that observed in densely ionizing radiations.

In India a wide range of physical and chemical mutagens have been used by several investigators for inducing mutations in different crop plants. Induction of mutations in wheat was first reported by Rajan (1940) followed by Sabnis and Mehta (1942). The effect of gamma rays was studied by Satyamurthy (1983).

Though the chemicals formed potential mutagens, the number of mutants released after their treatment is less as compared to those released from physical mutagenic treatments.

Tanaka (1969) treated 1405 genotypes of *Oryza sativa* by gamma rays and found that a number of genotypes showed improvement in regard to yield.

Anwar *et al.* (1993) induced mutations in safflower making use of gamma rays, EMS, NEU and sodium azide, and succeeded in recording a good amount of variability pertaining to a complex trait like quality/quantity of oil.

Ramana Rao *et al.* (1991) induced mutations in chilli using gamma rays and EMS. They studied the mutagenic effect on the quantitative characters like plant height, number of branches per plant and flowering.

Waghmare (1995) induced mutations for improving adaptability and yield attributes in *Gossypium barbadense* cv suvin and *Gossypium arboreum* cv Y_1. He recorded higher range of variability for number of monopodia, sympodia, single ball weight and germination percentage.

It has been realized that the mutation experiments offer a valuable tool in providing genetic variability and producing specific mutation that may confer a desired character in an otherwise superior variety, besides leading to an understanding of the mutation process. The direct use of induced mutation is a valuable supplementary approach to plant breeding. Mutation research has enabled the breeders to enlarge genetically conditioned variability of a species within a short time. In mutation breeding the normal processes of evolution are only speeded up and directed.

The usefulness of induced mutations in plant improvement rests on increasing the efficiency of mutation induction and its selection. The choice of proper mutagens and appropriate treatment conditions are evidently important in obtaining the desired efficiency and mutation rate.

Induced Mutation Studies in Okra (*Abelmoschus esculentus* (L.) Moench)

Mutagenesis has comprised an important method for okra breeding. This method has been employed to create variability and to widen the genetic base in okra (Jehangir and Chandrasekaran 1978).

Jambhle and Nerkar (1980) reported seed treatments of Pusa Sawani and Vaisali Vadhu varieties of okra with gamma rays and EMS. The induced chlorophyll mutations like *albina, chlorina* and M_2 viable mutations like dwarf plants with pointed fruits, palmatisect and crinckled leaf could be evidently noticed by them. The frequency of chlorophyll and viable mutation obtained in gamma ray treatment was found to be better than the EMS treatment.

Gonge and Kale (1993) determined the effect of gamma rays on plant height, internodal length, fruit length, fruit diameter, fruit yield and seed yield in section-2 cultivar of okra and recorded increased fruit yield, seed yield, decreased plant height, internodal length and fruit diameter in M_2 generation of that plant.

Jehangir and Chandrasekaran (1978) reported the effect of gamma rays and DES on fruit length, seed fertility, chlorophyll and viable mutants in M_2 generation. The gamma ray treated material showed chlorophyll mutation and increased fruit length/number of fruits, comprising an important achievement of okra mutagenesis programme.

Gregory (1956), Swaminathan (1965) and Gaul (1965) remarked that induced mutations are important in increasing genetic variability in many polygenic traits.

Senthil Kumar and Natarajan (1970) treated seeds of Pusa Sawani and Parbhani Kranti of okra by gamma rays and noted increased number of fruits per plant, seeds per fruit and seed weight in Pusa Sawani than Parbhani Kranti.

Shakeel Bhasha and Gopala Rao (1989) reported induced mutations for leaf area and physiological parameters like protein and chlorophyll content in M_2, after X-rays treatment in okra.

Abraham and Bhatia (1962) reported seeds of okra with EMS and in M_2 noticed the leaf curl and fruit thickness in Pusa Sawani cultivar of okra.

Kuwada (1967) reported seed treatment with gamma rays and X-rays and observed resistance to diseases and virus with an increasing dose of the pertinent physical mutagenic treatment.

The above brief review of related literature clearly indicates that despite multifaceted economic importance of *Abelmoschus esculentus* (L.) Moench, it has received rather limited attention of the scientists of our own country and very meager cytogenetical and mutational information is available from the laboratories of Indian researchers.

Hence it was considered very much relevant and appropriate to concentrate mainly on the aspect of mutation breeding in an economically important system like okra (*Abelmoschus esculentus* (L.) Moench) by using physical mutagen gamma rays and chemical mutagens ethyl methanesulphonate (EMS) and sodium azide (SA) so as to arrive at better understanding in regard to following aspects:

1. The biological effects of the physical and chemical mutagens in two varieties of okra *viz*. Parbhani Kranti and Arka Anamika.
2. The differential varietal response of okra to radiations and chemical mutagens.
3. The effectiveness and efficiency of the mutagens
4. The induction of variability in various quantitative characters and screening of micro-mutants.
5. Selection of macro-mutants having desirable attributes, and
6. The biochemical studies of selected mutants with reference to protein content.

Chapter 3

Materials and Methods for Induced Mutation

Materials

The seed material of two varieties of okra (*Abelmoschus esculentus* L. Moench.) namely Parbhani Kranti and Arka Anamika obtained from the Marathwada Agricultural University, Parbhani was used in the present study. The salient features of these two varieties of okra are given in the Table 5 (Figure 1).

Table 5: Salient Features of Two Varieties of Okra

Sl.No.	*Characteristics*	*Parbhani Kranti*	*Arka Anamika*
1.	Maturation duration	120	120-125
2.	Plant height (cm)	110-120	55-100
3.	Pod length (cm)	8-12	10-15
4.	Hundred seed weight (gm)	6.6	6.2
5.	Pod yield (q/h)	115	115
6.	Year of release	1985	1984
7.	Centre of release	Marathwada Agricultural University, Parbhani	IIHR Bangalore
8.	Other features	Tall, single stemmed, Particularly suitable for Marathwada and Karnataka.	Branched stem, Petiole and leaves are sparsely hairy suitable for all states.

q/h = quintle/hectare

Figure 1: The Field View of Okra Population.

Mutagens Used

In the present investigation the above mentioned two varieties of okra were treated by three mutagens. *viz.* physical mutagen, gamma rays and two chemical mutagens namely Ethyl methanesulphonate (EMS) and Sodium azide (SA).

Gamma Rays

Electromagnetic ionizing radiations were applied from a CO^{60} 1000 curie source of the gamma irradiation unit of the Department of Chemistry, Nagpur University, Nagpur. The dose rate was 24.578 rads per hour.

Ethyl methanesulphonate

Ethyl methanesulphonate ($CH_3SO_2OC_2H_5$) a monofunctional alkylating agent with a molecular weight of 124 manufactured by Koch light Laboratories Ltd. England was used in the present investigation.

Sodium Azide

This chemical mutagenic agent obtained from Sigma Chemical Company Ltd, U.S.A. was used in the present work.

Mode of Treatment with Mutagenic Agents

Gamma Rays

Healthy and dry seeds of the varieties Parbhani Kranti and Arka Anamika of okra having uniform size and equilibrated to the moisture level of 12 per cent were packed in small polythene covers and sealed for physical mutagen administration. The seed samples were exposed to doses of 5kR, 10kR and 20kR of gamma rays at the dose rate mentioned previously.

Chemical Mutagen

The begin with pilot experiments were conducted for determining the suitable concentration for further studies.

Preparation of Solution

The chemical mutagenic treatments were administered at room temperature of 25 ± 2°C. The fresh aqueous solutions of mutagens were prepared prior to treatment.

Treatment

The seeds were immersed in distilled water for seven hours to initiate presoaking. Such presoaked seeds were later on immersed in the mutagen solution for 5 hours with continuous shaking.

The presoaking enhances the rate of uptake of mutagen through increase in cell permeability and also initiates metabolism in the seed for treatment. The volume of chemical mutagenic solution used was 5-times as that of the seeds so as to facilitate uniform absorption. The seeds soaked in distilled water for 12-hours servred as control.

The different concentrations used for chemical mutagenic treatments were 0.05 per cent, 0.10 per cent and 0.15 per cent for EMS and 0.01 per cent, 0.02 per cent and 0.03 per cent for SA, respectively.

Immediately after the completion of treatment, the seeds were washed thoroughly under running tap water. Later on they were subjected to post-soaking in distilled water for one hour.

Raising of M_1-Generation

For each treatment, a batch of 400 presoaked seeds was used.100 seeds from each treatment were dried between the folds of filter paper and germinated in petriplates for recording germination percentage. The remaining lot of 300 seeds from each treatment was sown in field following randomised block design (RBD) with three replications along with controls for raising the M_1-generation. The seeds were sown at a distance of 30cm between plants and 45cm between rows. Individual plant was bagged to avoid cross-pollination.

Observation in M_1 Generation

Germination Percentage

The number of seeds showing emergence of the radical was counted from the seeds kept in petriplates lined with moist filter paper and expressed as percentage.

Plant Survival

The number of plants reaching maturity in the field was noted and expressed as percentage survival.

Pollen Sterility

Pollen sterility was determined from 10 randomly selected plants belonging to each treatment. The pollen grains from freshly dehisced anthers were stained with 1 per cent acetocarmine. Pollen grains that stained fully were counted as fertile while the empty partially stained and shrivelled ones were considered as sterile.

Studies in M_2 and M_3-Generations

Seeds of 25 normal looking plants of M_1 selected at random were collected on individual plant basis from all treatments and control. They were used for raising the M_2-generation on plant to a row basis.

The following observations were recorded in M_2-generation.

Frequency and spectrum of chlorophyll and viable mutations, calculation of mutagenic effectiveness and efficiency values based on chlorophyll mutation frequency and a critical assessment of induced variation in different quantitative characters.

Chlorophyll Mutations

The chlorophyll mutations were scored in the field when seedlings were 10-15 days old. The types of chlorophyll mutations scored were *xantha, chlorina* in different treatments.

The above spectrum of chlorophyll mutants was classified according to the terminology proposed by Gustafsson (1940) and Blixt (1961). The frequency of chlorophyll mutants was calculated according to Gaul (1957) *i.e.* number of mutants/100 M_2 plants.

Estimation of Mutagenic Effectiveness and Efficiency

Mutagenic effectiveness and efficiency of different mutagens were calculated according to the formulae suggested by Konzak *et al.* (1965).

Mutagenic effectiveness is a measure of the frequency of mutations induced by a unit dose of mutagen (kR, time × concentration) while mutagenic efficiency gives an idea of damage such as lethality, seedlng, injury, pollen sterility and chromosomal aberrations.

$$\text{Mutagenic effectiveness} = \frac{\text{Factor mutation (MF)}}{\text{Biological damage}}$$

$$= \text{MF/L, MF/I, MF/S,}$$

where,

MF: Per cent of chlorophyll mutations in M_2-generation.

L: Per cent of lethality in M_1-generation.

I: Per cent of seedling injury in M_1-generation.

S: Per cent of pollen sterility in M_1-generation.

Mutation Rate

The mutation rate was calculated by following formula,

$$\text{Mutation rate} = \frac{\text{Some of values of effectiveness or efficiency of particular Mutagen}}{\text{Number of treatments of the particular mutagen}}$$

This gives an idea of mutation induced by particular mutagen irrespective of dose.

Data on Quantitative Characters

From each treatment 25 plants were randomly selected for recording data on different quantitative characters in both M_2 and M_3 generations. Similarly 25 plants were picked up from controls for comparative assessment.

Data on the following quantitative characters were recorded.

1. Days to First Flowering

The date of first flowering of the plants was recorded from which the number of days required for the opening of the first flower and average number of days to flowering were recorded.

2. Days to First Picking

The data on number of days from sowing to first harvest of green marketable pods were recorded.

3. Number of Nodes on Main Stem

Total number of nodes from the base of plant to tip of main shoot was counted at the time of last picking.

4. Number of Pods per Plant

Number of green pods obtained from five plants were counted and an average number of green pods per plant was calculated.

5. Pod Length

Pod length was measured in centimeters from tip of pod to base of the calyx.

6. Green Pod Yield per Plant

Yield of green pods of five plants at each picking was taken. The yield of each picking was summed for total number of picking and the average of five plants was calculated and taken as yield per plant in grams.

7. Number of Seeds per Pod

Numbers of seeds in five random pods were counted and average number of seeds per pod was calculated.

8. Plant Height

The height of plant was recorded in centimeters from soil surface to the tip of main shoot at the time of harvesting.

9. 100 Seed Weight

The weight of hundred seeds was taken from five pods of each treatment.

10. First Flower Appearing Node

The number of nodes on which first flower appeared were recorded from five plants of each treatment.

11. Days to Maturity

The days to maturity of plants of all treatments and control was recorded from which the number of days required for maturation of the plant from the date of sowing was estimated.

Scoring of Variants in M_1 and Mutants in M_2/M_3 Generations

In M_1 Generation

The various leaf abnormalities and chlorophyll chimeras were noted. The frequency of M_1 plants carrying chlorophyll chimeras of different types (*viz. albina, xantha, chlorina,* and *viridis*) was calculated mutagen wise and variety wise, separately. The seeds obtained from these plants were sown separately for observation in M_2 generation.

In M_2 and M_3 Generations

Macromutations (Viable Mutations)

Mutations that can be scored externally and which affect the morphological characters of the plants were considered as macromutations. They were scored during the entire life cycle of the plant in M_2 and M_3 generations. All important details regarding flowering, fruiting and seed characters of such viable mutants were recorded.

The Flower/Fruit/Seed Mutations

A range of flower colour/fruit/seed mutations could be observed in the M_2 and M_3 generations of okra. The frequency of plants carrying all such mutations has been calculated on the basis of per 100 M_2 plants,

Micromutations

These are the mutations, which cannot be detected visually, but need biometrical analysis for their detection. Such plants were subjected to biometrical analysis and screened accordingly. The parameters were the same for M_2 and M_3 generations.

Statistical Analysis

The data were subjected to statistical and biometrical analysis for detecting the micromutations as given below.

The statistical model used for randomised block design comprised.

yijk = u + gij + bk + eijk

where,

yijk: Phenotypic performance of ijth mutations in kth block,
u: General mean,
gij: The effect of ijth mutations,
bk: The effect of kth block and
eijk: The environmental effect.

The analysis of variance based on this model is given below.

Source of Variation	*d. f.*	*Sum of Squares*	*Mean Squares Observed/Expected*	
Replications	r-1	Ssr	Msr	
Treatments	t-1	Sst	MSt	$\sigma^2e + r\sigma^2t$
Error	(r-1)(t-1)	Sse	Mse	σ^2e

where,

r: Number of replications and
t: Number of treatments.

The Mean squares were tested against error variance by usual 'f' test. The standard error of difference for comparing any two progeny means was estimated by formula.

SE(diff) = 2Mse/r

The critical difference was computed by multiplying the standard error of difference with 't' test values for (r-1)(t-1) degrees of freedom at 5 percent and 1 percent levels.

Various statistical data were calculated by using the following formulae.

Mean = $\Sigma x/n$

Variance = $(\Sigma x^2/n) - x^2$

Standard deviation (SD) = $\sqrt{\text{Variance}}$

Standard Error (SE) = $SD/\sqrt{N}$

Coefficient of variation (CV) = $(SD/\text{Mean}) \times 100$.

Chapter 4

Results and Findings of Induced Mutation Breeding

Mutagenic Studies

The results and discussion of the present study regarding mutagenic studies have been dealt with separately in regard to M_1, M_2 and M_3 generations respectively.

In M_1 generation studies were carried out on the differential effects of different mutagens on biological parameters and morphological abnormalities. In M_2 generation grown from the harvested seeds of M_1, studies were organized pertaining to chlorophyll mutations especially their frequency and spectrum, effectiveness and efficiency of the mutagens used besides screening and analysis of the quantitative characters, and several viable macro mutants which were selected from M_2 population have also been and reported.

In M_3 generation grown from the harvested seeds of M_2, the study included screening and analysis of quantitative characters and field testing of the viable macro-mutants isolated from M_2 population regarding their yield characteristics and biochemical features.

Studies in M_1 Generation

Biological Parameters

The present investigation comprised the study of mutagenic sensitivity of two varieties (Parbhani Kranti and Arka Anamika) of okra in regard to some biological/ morphological parameters. The effect of gamma rays, EMS and SA on okra was studied by choosing three concentrations for each mutagen (Gamma rays 5 kR, 10 kR and 20 kR and EMS 0.05 per cent, 0.10 per cent and 0.15 per cent and SA 0.01 per

Figure 2: Plant Habit of Variety Parbhani Kranti.

Figure 3: Plant Habit of Variety Arka Anamika.

cent, 0.02 per cent and 0.03 per cent). The results have been organized separately for each parameter.

The parameters selected for M_1 generation comprised

1. Seed germination and lethality
2. Pollen sterility
3. Plant survival at maturity
4. Chlorophyll chimeras
5. Morphological abnormalities.

Mutagenic Sensitivity

Seed Germination (Lethality) (Tables 6, 7) (Graph 1)

In okra the maximum number of seeds germinated on 6th day after sowing in both the varieties, namely Parbhani Kranti and Arka Anamika. In control the germination percentage was found to be 88.00 per cent in Parbhani Kranti and 84.32 per cent in Arka Anamika, respectively.

After the mutagenic treatments, an inhibitory effect on seed germination could be distinctly seen in variety Parbhani Kranti. In variety Arka Anamika on the other hand a mixed pattern of stimulation/inhibition could be noticed in the gamma ray, EMS and SA treatments.

Table 6: The Effect of Mutagens on Seed Germination in *Abelmoschus esculentus* (L.) Moench.

Variety : Parbhani Kranti

Mutagen	*Dose*	*Germination (Per cent)*
Control	Control	88.00
Gamma rays	5kR	60.00
	10kR	56.22
	20kR	52.00
EMS (per cent)	0.05	72.20
	0.10	69.20
	0.15	70.44
SA (per cent)	0.01	78.32
	0.02	72.01
	0.03	71.65

± SE = 3.35

Table 7: The Effect of Mutagens on Seed Germination in *Abelmoschus esculentus* (L.) Moench.

Variety : Arka Anamika

Mutagen	*Dose*	*Germination (Per cent)*
Control	–	84.32
Gamma rays	5kR	66.40
	10kR	63.00
	20kR	62.40
EMS (per cent)	0.05	82.40
	0.10	86.03
	0.15	81.22
SA (per cent)	0.01	88.21
	0.02	82.43
	0.03	86.20

± SE = 3.22

Gamma ray treatment revealed a gradual decreasing trend in germination from lower to higher doses in both the varieties. The germination ranged from 60.00 per cent to 52.00 per cent in variety Parbhani Kranti and 66.40 per cent to 62.40 per cent in variety Arka Anamika of okra.

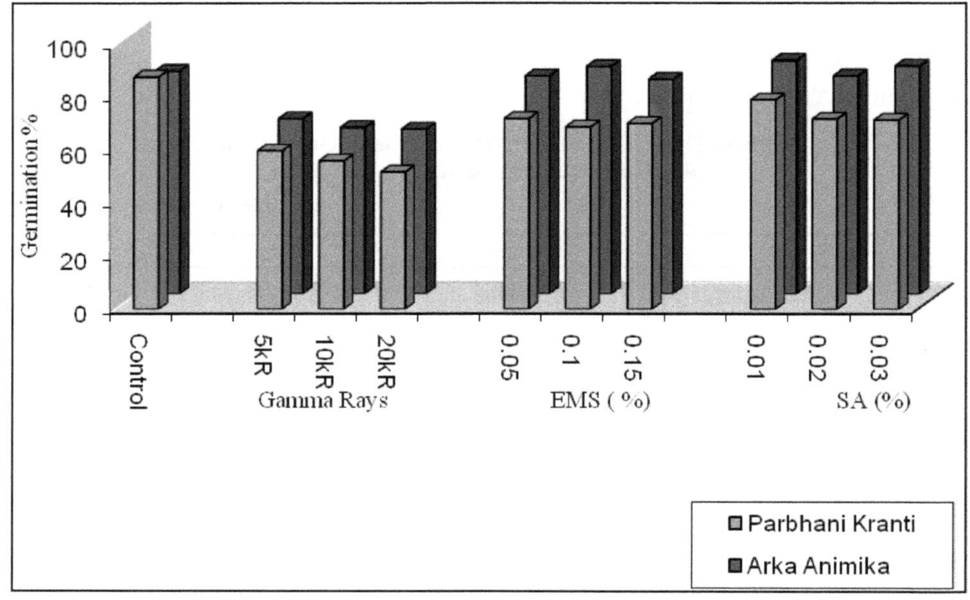

Graph 1: Effect of Mutagens on Seed Germination in *Abelmoschus esculentus* L. Moench.

After EMS treatment in variety Parbhani Kranti the germination indicated an inhibitory feature at all the three concentrations of the mutagen. In variety Arka Anamika except middle concentration (0.10 per cent) at the remaining two concentrations (0.05 per cent and 0.15 per cent), the germination percentage showed inhibitory effect and ranged from 69.20 per cent to 72.20 per cent in Parbhani Kranti and 81.22 per cent to 86.03 per cent in Arka Anamika varieties.

The SA treatment resulted in reducing seed germination percentage at all its concentrations in variety Parbhani Kranti. In contrast to this except for the middle concentration (0.02 per cent) at the remaining two concentrations (0.01 per cent and 0.03 per cent), the germination percentage showed stimulation over control in variety Arka Anamika. The germination ranged from 71.65 to78.32 per cent in variety Parbhani Kranti and 82.43 per cent to 88.21 per cent in variety Arka Anamika after SA treatments.

Survival Percentage at Maturity (Tables 8, 9) (Graph 2)

The survival of plants at maturity revealed wide fluctuations in their values with dose/concentration of all the mutagens in variety Parbhani Kranti. The values ranged from 79.63 per cent to 85.23 per cent in gamma rays. 89.03 per cent to 91.64 per cent in EMS and 85.53 per cent to 92.24 per cent in SA treatments.

Table 8: The Effect of Mutagens on Survival of Plants at Maturity in *Abelmoschus esculentus* (L.) Moench.

Variety : Parbhani Kranti

Mutagen	*Dose*	*Survival of Plants at Maturity (per cent)*
Control	–	89.04
Gamma rays	5kR	85.23
	10kR	82.42
	20kR	79.63
EMS (per cent)	0.05	91.22
	0.10	89.03
	0.15	91.64
SA (per cent)	0.01	83.53
	0.02	92.24
	0.03	88.66

± SE = 1.36

Table 9: The Effect of Mutagens on Survival of Plants at Maturity in *Abelmoschus esculentus* (L.) Moench.

Variety : Arka-Anamika.

Mutagen	*Dose*	*Survival of Plants at Maturity (per cent)*
Control	–	82.54
Gamma rays	5kR	72.40
	10kR	74.43
	20kR	69.02
EMS (per cent)	0.05	83. 67
	0.10	82.05
	0.15	87.33
SA (per cent)	0.01	87.43
	0.02	89.57
	0.03	83.03

± SE = 2.20

In case of variety Arka Anamika, values of survival of plants increased in EMS and SA treatments. In Gamma ray treatment inhibitory effect could be seen. The values ranged from 69.02 per cent to 72.40 per cent in gamma rays, 82.05 per cent to 87.33 per cent in EMS and 83.03 per cent to 89.57 per cent in SA treatment of variety Arka Anamika.

The highest survival values (92.24 per cent) was recorded at 0.02 per cent SA in variety Parbhani Kranti. The lowest survival percentage (72.40 per cent) could be seen at 5 kR gamma ray treatment in variety Arka Anamika.

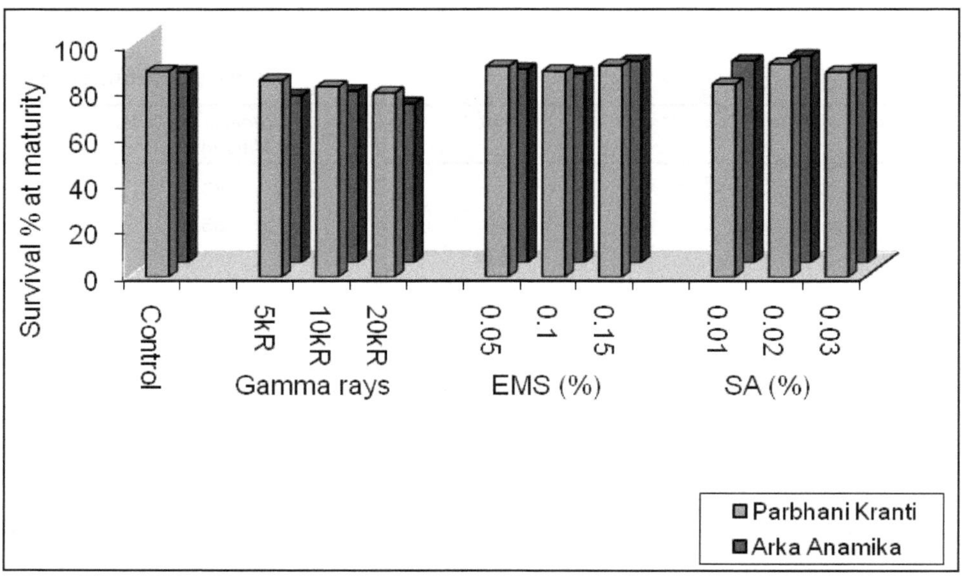

Graph 2: Effect of Mutagens on Survival of Plants at Maturity in *Abelmoschus esculentus* L. Moench.

Leaf Morphological Changes (Tables 10, 11) (Graph 3)

In all treatments it was observed that the leaves of the plants exhibited various shapes and sizes. The leaf variations comprised (i) suppression of leaf lamina, and (ii) reduction in size of leaves.

Table 10: The Effect of Mutagens on the Frequency of Plant Carrying Leaf Morphological Changes in *Abelmoschus esculentus* (L.) Moench.

Variety : Parbhani Kranti

Mutagen	*Dose*	*Frequency of Plant Carrying Leaf Morphological Changes (per cent)*
Control	–	00
Gamma rays	5kR	7.58
	10kR	12.03
	20kR	15.42
EMS (per cent)	0.05	5.43
	0.10	11.14
	0.15	10.38
SA (per cent)	0.01	7.02
	0.02	11.36
	0.03	10.00

± SE = 1.03

Table 11: The Effect of Mutagens on the Frequency of Plant Carrying Leaf Morphological Changes in *Abelmoschus esculentus* (L.) Moench.

Variety : Arka Anamika

Mutagen	*Dose*	*Frequency of Plant Carrying Leaf Morphological Changes (per cent)*
Control	–	–
Gamma rays	5kR	6.34
	10kR	11.12
	20kR	16.21
EMS (per cent)	0.05	3.97
	0.10	1.28
	0.15	5.03
SA (per cent)	0.01	6.42
	0.02	5.51
	0.03	7.78

± SE = 1.45

The frequency of plants carrying the leaf morphological changes varied from dose/concentration in different mutagens used. Of the three mutagens the gamma rays, especially its 20kR dose succeeded in inducing the highest frequency (16.21 per cent and 15.42 per cent) of the plants with leaf abnormalities in Arka Anamika and Parbhani Kranti. The lowest frequency of leaf changes carrying plants (5.43

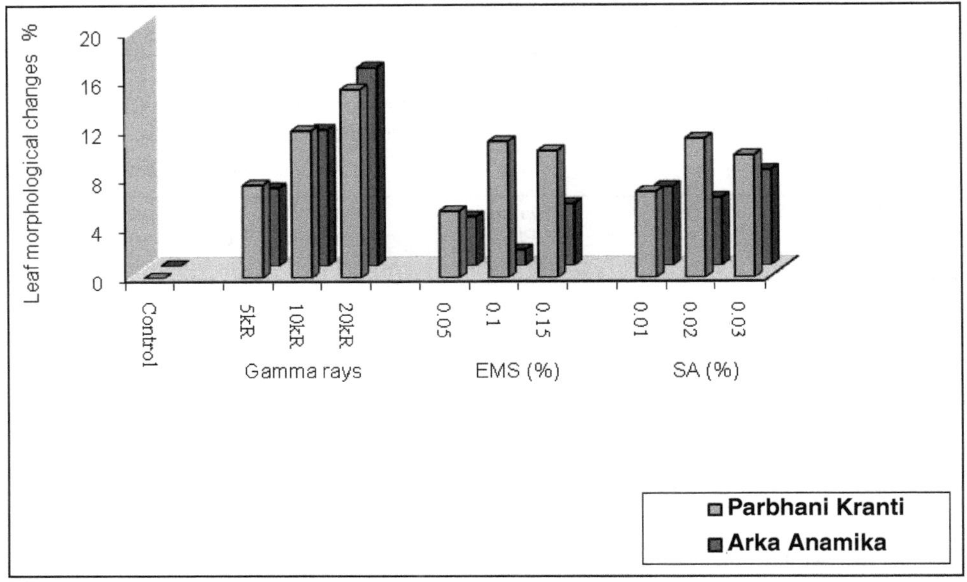

Graph 3: Effect of Mutagens on Frequency of Plants Carrying Leaf Morphological Changes in *Abelmoschus esculentus* L. Moench.

per cent and 1.28 per cent) could be noted at 0.05 per cent EMS in variety Parbhani Kranti and at 0.10 per cent EMS in variety Arka Anamika.

Chlorophyll Deficient Sectors (Tables 12, 13) (Figures 4 and 5) (Graph 4)

The critical screening of the M_1 population of okra demonstrated the induction of chlorophyll chimeras of different types in the leaves at all the mutagenic treatments. These chimeras were at yellow green (*Chlorina*), yellow (*Xantha*), white (*Albina*) and light green (*viridis*) types, and they were located at the margins of leaves or distributed almost throughout the leaf lamina.

Table 12: The Effect of Mutagens on the Frequency of Plants Carrying Chlorophyll Deficient Chimeras in *Abelmoschus esculentus* (L.) Moench.

Variety : Parbhani Kranti

Mutagen	*Dose*	*Number of Plants Observed*	*Number of Plants with Chimeras*	*Frequency of Chlorophyll Deficient Sectorial Plants*
Control	–	–	–	–
Gamma rays	5kR	158	3	1.8
	10kR	182	2	1.09
	20kR	114	3	2.60
EMS (per cent)	0.05	131	2	1.52
	0.10	166	1	0.60
	0.15	142	–	–
SA (per cent)	0.01	123	3	2.43
	0.02	174	2	1.14
	0.03	161	5	3.10

Table 13: The Effect of Mutagens on the Frequency of Plants Carrying Chlorophyll Deficient Chimeras in *Abelmoschus esculentus* (L.) Moench.

Variety : Arka Anamika

Mutagen	*Dose*	*Number of Plants Observed*	*Number of Plants with Chimeras*	*Frequency of Chlorophyll Deficient Sectorial Plants*
Control	–	–	–	–
Gamma rays	5KR	128	2	1.56
	10kR	152	1	0.65
	20kR	191	1	0.52
EMS (per cent)	0.05	116	2	1.72
	0.10	163	–	–
	0.15	159	3	1.88
SA (per cent)	0.01	138	4	2.89
	0.02	167	1	0.59
	0.03	110	5	4.54

Figure 4: The M_1 Plant of Okra Showing *Albina* Chlorophyll Deficient Sector.

Figure 5: The M_1 Plant of Okra Showing *Xantha* Chlorophyll Deficient Sector.

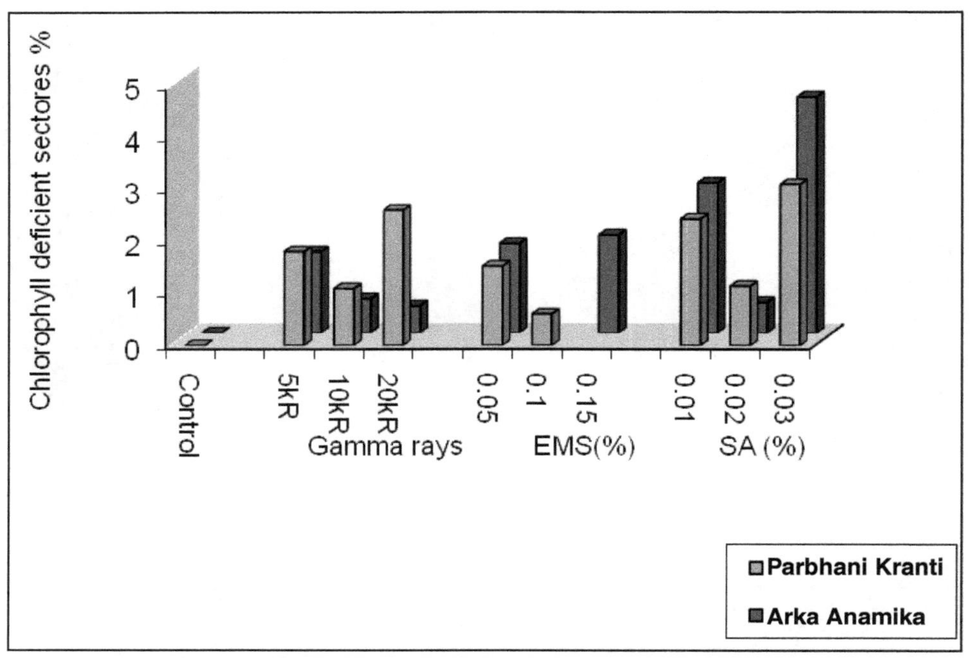

Graph 4: Effect of Mutagens on Frequency of Plants Carrying Chlorophyll Chimeras in *Abelmoschus esculentus* L. Moench.

They were noticeable at a higher frequency in the material treated with SA, while the material treated with EMS and Gamma rays indicated relatively lower frequency values for chimera carrying plants in both the varieties of okra (Parbhani Kranti and Arka Anamika).

The higher frequency values (3.10 per cent and 4.24 per cent) were noticeable at (0.03 per cent) SA treatments of variety Parbhani Kranti and Arka Anamika, respectively.

Pollen Sterility (Tables 14, 15) (Graph 5)

The pollen sterility in control plants was 1.05 per cent and 1.09 per cent in Parbhani Kranti and Arka Anamika, respectively. The maximum pollen sterility (10.50 per cent) could be seen at 5kR dose of gamma rays in Parbhani Kranti, while in Arka Anamika the same feature was demonstrated by the 5kR gamma ray treatment, indicating the sterility value of 10.79 per cent. Of the three mutagens it was EMS, which induced the least pollen sterility in variety Parbhani Kranti, while in Arka Anamika it was SA which induced least pollen sterility. The pollen sterility values ranged from 6.23 per cent to 10.5 per cent and 4.02 per cent to 10.79 per cent after gamma ray, 2.31 per cent to 3.11 per cent and 2.68 per cent to 4.46 per cent after EMS and 4.33 per cent to 6.26 per cent and 2.42 per cent to 4.07 per cent after SA treatments in varieties Parbhani Kranti and Arka Anamika, respectively.

Table 14: The Effect of Mutagens on Pollen Sterility in *Abelmoschus esculentus* (L.) Moench.

Variety : Parbhani Kranti

Mutagen	*Dose*	*Pollen Sterility (Per cent)*
Control	–	1.05
Gamma rays	5kR	10.5
	10kR	6.23
	20kR	8.33
EMS (per cent)	0.05	2.31
	0.10	2.51
	0.15	3.11
SA (per cent)	0.01	4.33
	0.02	5.12
	0.03	6.26

± SE = 0.94

Table 15: The Effect of Mutagens on Pollen Sterility in *Abelmoschus esculentus* (L.) Moench.

Variety : Arka Anamika

Mutagen	*Dose*	*Pollen Sterility (Per cent)*
Control	–	1.09
Gamma rays	5kR	10.79
	10kR	7.12
	20kR	4.02
EMS (per cent)	0.05	2.68
	0.10	3.32
	0.15	4.48
SA (per cent)	0.01	2.42
	0.02	3.68
	0.03	4.07

± SE = 0.88

Studies in M_2 and M_3 Generations

Chlorophyll Mutations (Tables 16-19) (Graph 6)

The M_2 generation was raised from the seed progenies of M_1 plants. The chlorophyll mutants were scored at the seedling stage. They were of different types such as *xantha, chlorina, chloroxantha* and *viridis* (Figures 6-8).

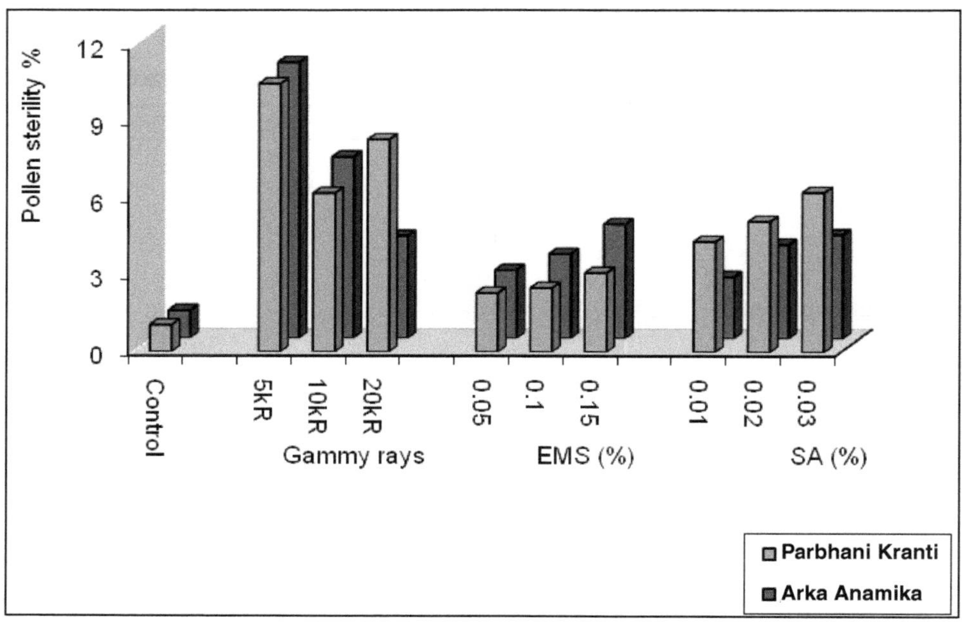

Graph 5: Effect of Mutagens on Pollen Sterility in *Abelmoschus esculentus* L. Moench.

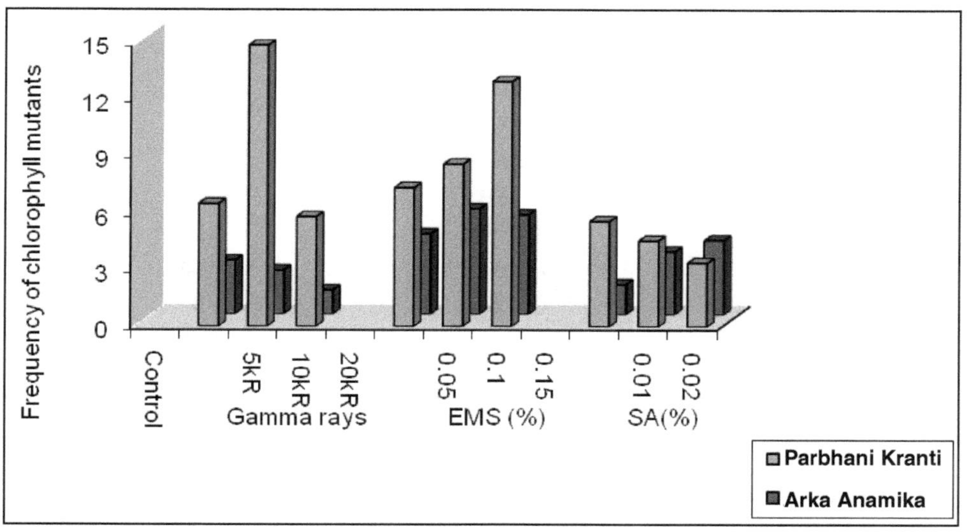

Graph 6: Effect of Mutagens on Chlorophyll Mutants in *Abelmoschus esculentus* L. Moench.

Xantha mutants displayed a bright yellow colour. In some mutants the colour tinge was somewhat lighter. These mutants survived for nearly 40 to 50 days and demonstrated stunted growth pattern during progressive growth stages.

Figure 6: The *Viridis* Mutant in M_2 Generation of Okra.

Figure 7: The *Xantha* Mutant in M_2 Generation of Okra.

The colour of *Chlorina* mutants was yellowish green. A few of them reverted to normal green type.

The *viridis* mutants showed dull light green colour. This colour gradually changed to the normal green colour during the subsequent growth phases of the plant.

Figure 8: The *Chlorina* Mutant in M_2 Generation of Okra.

The *xantha* type mutants revealed yellowish colour pattern and quite a few of them survived and developed branches and leaves. They however showed delayed pod formation with less number of seeds per pod. Most of them bred true in the subsequent M_3 generation.

Table 16: The Effect of Mutagens on Frequency of Chlorophyll Mutants in M_2 Generation in *Abelmoschus esculentus* (L.) Moench.

Variety : Parbhani Kranti

Mutagen	*Dose*	*Number of Plants Screened*	*Total Number of Chlorophyll Mutants*	*Frequency of Chlorophyll Mutants*
Control	–	480	–	–
Gamma rays	5kR	368	24	6.52
	10kR	235	35	14.89
	20kR	325	19	5.84
EMS (per cent)	0.05	340	25	7.35
	0.10	408	35	8.57
	0.15	270	35	12.96
SA (per cent)	0.01	125	7	5.60
	0.02	175	8	4.57
	0.03	180	6	3.38

Table 17: The Effect of Mutagens on Frequency of Chlorophyll Mutants in M_2 Generation in *Abelmoschus esculentus* (L.) Moench.

Variety : Arka Anamika

Mutagen	*Dose*	*Number of Plants Screened*	*Total Number of Chlorophyll Mutants*	*Frequency of Chlorophyll Mutants*
Control	–	520	–	–
Gamma rays	5kR	523	15	2.86
	10kR	450	6	2.33
	20kR	312	4	1.28
EMS (per cent)	0.05	210	9	4.28
	0.10	125	7	5.60
	0.15	170	9	5.29
SA (per cent)	0.01	312	5	1.60
	0.02	270	9	3.33
	0.03	304	12	3.94

The *chlorina* plant revealed dull light green colour and most of them could not survive because of retarded vegetative growth.

In case of okra, all the three mutagens induced different types of chlorophyll mutants. The spectrum of induced chlorophyll mutants was quite broad in both the varieties of okra, and the response of the varieties towards the different mutagens was differential.

Table 18: The Effect of Mutagens on the Spectrum of Chlorophyll Mutants in M_2 Generation in *Abelmoschus esculentus* (L.) Moench.

Variety : Parbhani Kranti

Mutagen	*Dose*	*Frequency of Chlorophyll Mutants*	*Relative Percentage*		
			Xantha	*Chlorina*	*Viridis*
Control	–	–	–	–	–
Gamma rays	5kR	6.52	–	48.05	51.94
	10kR	14.89	2.82	51.25	45.92
	20kR	5.14	67.41	30.12	2.44
EMS (per cent)	0.05	7.35	–	25.18	74.66
	0.10	8.57	8.56	58.33	32.10
	0.15	12.96	12.20	39.24	47.56
SA (per cent)	0.01	5.60	14.18	36.16	47.66
	0.02	4.57	4.22	30.08	64.77
	0.03	3.33	–	14.92	84.22

Table 19: The Effect of Mutagens on the Spectrum of Chlorophyll Mutants in M_2 Generation in *Abelmoschus esculentus* (L.) Moench.

Variety : Arka Anamika

Mutagen	*Dose*	*Frequency of Chlorophyll Mutants*	*Relative Percentage*		
			Xantha	*Chlorina*	*Viridis*
Control	–	–	–	–	–
Gamma rays	5kR	2.86	35.42	55.07	9.51
	10kR	1.33	32.12	42.16	25.70
	20kR	1.28	–	71.23	28.76
EMS (per cent)	0.05	4.28	9.12	73.28	17.55
	0.10	5.60	24.07	50.56	25.36
	0.15	5.29	–	34.98	65.02
SA (per cent)	0.01	1.60	5.9	45.8	65.00
	0.02	3.33	2.3	78.32	19.37
	0.03	3.94	–	90.81	9.18

In variety Parbhani Kranti the frequency of chlorophyll mutants increased linearly from 6.5 per cent to 7.8 per cent with gradual increase in gamma ray dose from 5kR to 10 kR and at 20kR it decreased (5.8 per cent). In variety Arka Anamika however, the frequency of chlorophyll mutants decreased linearly from 2.8 per cent to 1.20 per cent with gradual increase in gamma ray dose from 5kR to 20 kR.

Among the two chemical mutagens, EMS (0.15 per cent) proved to be very much successful in inducing the highest frequency (12.9 per cent) of chlorophyll mutants in variety Parbhani Kranti while in variety Arka Anamika, the EMS concentration (0.10 per cent) induced the highest frequency (5.6 per cent) of chlorophyll mutants.

The frequency of chlorophyll mutants varied from 5.8 per cent to 7.8 per cent after gamma ray, 7.3 per cent to 12.9 per cent after EMS and 3.3 per cent to 5.6 per cent after SA treatment in variety Parbhani Kranti. As regards to variety Arka Anamika however, the frequency of chlorophyll mutants varied from 1.2 per cent to 2.8 per cent after gamma ray treatment, 4.2 per cent to 5.6 per cent afer EMS treatment and 1.6 per cent to 3.9 per cent after SA treatment.

As regards the *xantha* mutants the 20kR dose of gamma rays proved to be the most effective in inducing the highest relative percentage (67.41 per cent) in variety Parbhani Kranti. While in variety Arka Anamika the highest relative percentage of *xantha* (35.42 per cent) could be observed at 10kR of gamma rays.

As regards the *chlorina* type of mutant it could display its highest relative percentage (58.33 per cent) in variety Parbhani Kranti at 0.10 per cent EMS treatment. The highest relative percentage (90.81 per cent) of *chlorina* could be observed at 0.03 per cent SA treatment, followed by 78.32 per cent at 0.02 per cent SA and 71.23 per cent at 20kR gamma ray treatment, in variety Arka Anamika.

The *viridis* type of chlorophyll mutant revealed highest relative percentage (84.22 per cent) in variety Parbhani Kranti at 0.03 per cent SA treatment, and lowest relative percentage (2.44 per cent) at 20kR gamma ray treatment. In variety Arka Anamika however the highest relative percentage (65.00 per cent) could be seen at 0.01 per cent SA treatment and the lowest relative percentage (9.18 per cent) could be noted at 0.03 per cent SA. The spectrum of chlorophyll mutants and the frequency of individual mutant type differed in different mutagens. The *albina* type of chlorophyll mutant was not observed in any of the varieties of okra.

Mutagenic Effectiveness (Tables 20, 21)

The mutagenic effectiveness is a measure of factor mutations induced by a unit dose of mutagen. The major trends pertaining to this parameter influenced by different mutagens can be understood through a critical perusal of tables 20-21.

In gamma ray treatments, there was decrease in effectiveness value with increasing doses in Parbhani Kranti and Arka Anamika, except for 10kR gamma ray in Parbhani Kranti, whereas in EMS treatment the effectiveness values decreased linearly with increasing concentrations in both varieties of okra, except for the 0.15 per cent EMS treatment in Parbhani Kranti.

In SA treatment the effectiveness values decreased with increasing concentrations in variety Parbhani Kranti and in Arka Anamika, except for 0.02 per cent treatment.

The highest mutagenic effectiveness of 140.00 was observable at 0.01 per cent concentration of SA in variety Parbhani Kranti while in Arka Anamika the highest value (41.62) could be recorded at 0.02 per cent SA concentration.

Table 20: The Effectiveness of Mutagens in the M_2 Generation in *Abelmoschus esculentus* (L.) Moench.

Variety : Parbhani Kranti

Mutagen	*Dose*	*Frequency of Chlorophyll Mutans*	*Effectiveness Mf/dose or Mf/Txc*
Control	–	–	–
Gamma rays	5kR	6.52	1.30
	10kR	14.89	1.48
	20kR	5.84	0.29
EMS (per cent)	0.05	7.35	36.75
	0.10	8.57	21.42
	0.15	12.96	21.60
SA (per cent)	0.01	5.60	140.00
	0.02	4.57	57.12
	0.03	3.33	27.75

Table 21: The Effectiveness of Mutagens in M_2 Generation in *Abelmoschus esculentus* (L.) Moench.

Variety : Arka Anamika

Mutagen	*Dose*	*Frequency of Chlorophyll Mutans*	*Effectiveness Mf/dose or Mf/Txc*
Control	–	–	–
Gamma rays	5kR	2.86	0.57
	10kR	1.33	0.139
	20kR	1.28	0.08
EMS (per cent)	0.05	4.28	21.4
	0.10	5.60	14.0
	0.15	5.29	8.81
SA (per cent)	0.01	1.60	40.00
	0.02	3.33	41.62
	0.03	3.94	32.83

Mutagenic Efficiency (Tables 22, 23)

Efficient mutagenesis is the proportion desirable changes (mutations) free from associated undesirable changes. The mutagenic efficiency is the ratio of chlorophyll mutations induced in M_{-2} generation to various biological damages induced in M_1 generation such as lethality and pollen sterility. The tables 22 and 23 present the data on efficiency of mutagens in relation to various biological effects indicated earlier.

As far as gamma rays are concerned, in variety Parbhani Kranti the lowest dose 5kR was found to be most efficient in regard to lethality and pollen sterility, but in variety Arka Anamika, the efficiency of mutagens in M_2 generation was found to be maximum in regard to lethality (0.085) and pollen sterility (0.365) at 5kR and 10kR dose of gamma rays. In variety Arka Anamika, the efficiency of mutagens decreased with an increase in dose in regard to lethality. In variety Parbhani Kranti, the efficiency of mutagens increased gradually with an increase in dose in regard to lethality and pollen sterility.

Among the chemical mutagens, the EMS showed lowest efficiency (0.264) at its 0.05 per cent concentration pertaining to lethality. It showed highest value (0.438) at 0.15 per cent EMS in variety Parbhani Kranti. In variety Arka Anamika the EMS treatment showed lowest efficiency (0.243) at 0.05 per cent in regard to lethality and showed the highest efficiency (0.400) at EMS 0.10 per cent. The efficiency of mutagens indicated lowest value (3.160) at 0.05 per cent EMS in regard to pollen sterility in Parbhani Kranti and showed the highest efficiency (4.147) at 0.15 per cent EMS. Whereas in variety Arka Anamika, the efficiency of mutagens demonstrated highest value (1.686) in regard to pollen sterility at 0.10 per cent EMS and it showed lowest efficiency (1.165) at 0.15 per cent EMS treatment.

In SA treatment, the efficiency values decreased with increasing concentrations in respect of pollen sterility in Parbhani Kranti, but in variety Arka Anamika efficiency values increased with increasing concentration in respect of pollen sterility. In regard to lethality the efficiency values decreased with the increasing concentration in respect to lethality in variety Parbhani Kranti. In variety Arka Anamika the efficiency values increased with increasing concentration in regard to lethality. In variety Arka Anamika the lowest efficiency (0.135) was at 0.01 per cent SA while the highest value (0.285) could be observed at 0.03 per cent of SA treatment.

Table 22: The Relative Efficiency of Mutagens in M_2 Generation of *Abelmoschus esculentus* (L.) Moench.

Variety : Parbhani Kranti

Mutagen *Control*	*Dose* *Control*	*Per cent Chlorophyll Mutants (Mf)*	*Per cent Lethality (L)*	*Efficiency Lethality Mf/L*	*Pollen Sterility (S)*	*Efficiency Mf/S*
Gamma rays	5kR	6.52	40.00	0.163	10.50	0.619
	10kR	14.89	43.78	0.340	6.23	1.252
	20kR	5.84	48.00	0.121	8.33	0.696
EMS (per cent)	0.05	7.35	27.80	0.264	2.31	3.160
	0.10	8.57	30.80	0.278	2.51	3.386
	0.15	12.96	29.56	0.438	3.11	4.147
SA (per cent)	0.01	5.60	21.68	0.258	4.33	1.293
	0.02	4.57	27.99	0.163	5.12	0.878
	0.03	3.33	28.35	0.117	6.26	0.527

Table 22: The Relative Efficiency of Mutagens in M_2 Generation of *Abelmoschus esculentus* (L.) Moench.

Variety : Arka Anamika

Mutagen *Control*	*Dose* *Control*	*Per cent Chlorophyll Mutants (Mf)*	*Per cent Lethality (L)*	*Efficiency Lethality Mf/L*	*Pollen Sterility (S)*	*Efficiency Mf/S*
Gamma rays	5kR	2.86	33.60	0.085	10.79	0.259
	10kR	1.33	37.00	0.070	7.12	0.365
	20kR	1.28	37.60	0.035	4.02	0.298
EMS (per cent)	0.05	4.28	17.60	0.243	2.68	1.567
	0.10	5.60	13.97	0.400	3.32	1.686
	0.15	5.29	18.78	0.281	4.46	1.165
SA (per cent)	0.01	1.60	11.79	0.135	2.42	0.661
	0.02	3.33	17.57	0.189	3.68	0.896
	0.03	3.94	13.80	0.285	4.07	0.958

In Parbhani Kranti, both the parameters *viz.* lethality and pollen sterility increased with gradual increase in concentration as regards the mutagenic efficiency in EMS treatment, is concerned.

In the variety Arka Anamika the efficiency values increased with increasing concentration of the mutagens in regard to lethality and pollen sterility. The highest efficiency value (1.686) could be seen at 0.10 per cent EMS concentration and the lowest efficiency (0.259) could be noted at 5kR gamma ray treatment in this variety.

Mutation Rate (Tables 24, 25)

The mutation rates were calculated taking into consideration the mean values of efficiency for each treatment. This has given an idea about the average rate of mutation induction per mutagen. The mutation rates of the mutagen based on efficiency are presented in Tables 24 and 25.

Table 24: The Mutation Rates of the Mutagens Based on Efficiency in M_2 Generation of *Abelmoschus esculentus* (L.) Moench.

Variety : Parbhani Kranti

Mutagen	*Mutation Rates Based on*	
	Lethality	*Sterility*
Control	–	–
Gamma rays	0.069	0.28
EMS (per cent)	0.108	1.180
SA (per cent)	0.059	0.29

Table 25: The Mutation Rates of the Mutagens Based on Efficiency in M_2 Generation of *Abelmoschus esculentus* (L.) Moench.

Variety : Arka Anamika

Mutagen	*Mutation Rates Based on*	
	Lethality	*Sterility*
Control	–	–
Gamma rays	0.021	0.102
EMS (per cent)	0.102	0.49
SA (per cent)	0.067	0.27

If we consider the mutation rates based on efficiency, the order of mutagens changes as mutagens have different values in relation to lethality and pollen sterility.

Taking into consideration the mutation rates for lethality, the values were 0.069 (gamma rays), 0.108 (EMS) and 0.059 (SA) in variety Parbhani Kranti. In variety Arka Anamika the values of mutation rates were 0.021 (gamma rays), 0.102 (EMS) and 0.067 (SA)

Thus with respect to lethality the order of mutagens in both the varieties has been

In variety Parbhani Kranti : SA < Gamma rays < EMS and

In variety Arka Anamika: Gamma rays < SA < EMS.

When the mutation rate for pollen sterility was considered, the values were 0.28 (gamma rays), 1.180 (EMS) and 0.29 (SA) in variety Parbhani Kranti. The values for Arka Anamika were 0.102 (gamma rays), 0.49 (EMS) and 0.27 (SA). In respect of the pollen sterility, the sequence of mutagens in both varieties was as follows:

In variety Parbhani Kranti : Gamma rays < SA < EMS and

In variety Arka Anamika:. SA < Gamma rays < EMS.

Viable Mutants (Tables 26, 27)

Quite a good number of viable mutants have been observed in M_2 generation of both the varieties of okra. The frequency of viable mutants revealed an increasing trend with gradual increase in dose/concentration of all the mutagens in variety Parbhani Kranti. In variety Arka Anamika the frequency of viable mutants increased with an increasing dose/concentration of EMS and SA treatments while it fluctuated in gamma ray treatment of variety Arka Anamika.

The chemical mutagens EMS and SA showed increasing trend with increasing concentrations in both varieties of okra *viz.* Parbhani Kranti and Arka Anamika. In both the varieties of the plant, the viable mutants obtained were of the following types. 1) Dwarf mutant 2) Tall mutant 3 Branched 4) Early flowering 5) Long pod 6) Early maturing 7) Late maturing 8) High yielding.

Description of Viable Mutants

I) Viable Mutants in Variety Parbhani Kranti (Table 28)

1) Dwarf Mutant

These mutants were characterized by an extreme reduction in plant height. It was 88.72 cm as against 108.36 cm for control. All the characters in this mutant demonstrated reduced values as compared with control.

2) Tall Mutant

The tall mutant showed a height of 136.28 cm as against 108.36 cm in control. These mutants showed positive correlation between height of the plant with nodes on main stem. The period for maturity in this mutant was comparable with that of control.

3) Branched Mutant

These mutants showed 4-5 branches. They exhibited reduction in nodes on branches and pod length (14.8 cm).

4) Early Flowering Mutant

These mutants attained flowering almost seven days earlier than the control population. They flowered in 36.37 days as against 43.62 days for control. These mutants showed positive correlation between days to flowering and 100 seed weight (6.10 gm).

Table 26: Effect of Mutagens on Frequency and Spectrum of Viable Mutants in M_2 Generation of *Abelmoschus esculentus* (L.) Moench.

Variety : Parbhani Kranti

Mutagens	*Concentration*	*Frequency of Viable Mutants*	*Relative Percentage*							
			Dwarf Mutant	*Tall Mutant*	*Branched Mutant*	*Early Flowering*	*Long Pod Mutant*	*Early Maturing*	*Late Maturing*	*High Yielding*
Control	Control	–	–	–	–	–	–	–	–	–
Gamma rays	5kR	10.15	1.17	1.49	1.97	1.29	1.29	0.78	0.96	1.20
	10kR	12.92	3.27	2.64	2.90	1.45	–	0.65	1.13	1.88
	20kR	13.90	1.84	1.08	–	2.41	1.26	1.58	3.21	2.52
EMS (per cent)	0.05	14.26	2.16	1.37	2.50	0.97	2.05	2.81	1.14	1.26
	0.10	15.01	0.62	2.81	1.43	3.16	1.10	2.98	2.03	0.88
	0.15	15.76	2.08	1.68	–	2.92	2.00	3.28	1.76	2.04
SA (per cent)	0.01	14.42	0.78	2.81	1.89	2.19	1.27	0.86	1.98	2.64
	0.02	16.52	4.16	–	2.53	2.77	1.18	2.52	2.26	1.10
	0.03	21.93	3.76	2.32	2.88	3.11	–	4.89	4.28	2.69

Table 27: Effect of Mutagens on Frequency and Spectrum of Viable Mutants in M_2 Generation of *Abelmoschus esculentus* (L.) Moench.

Variety : Arka Anamika

Mutagens	Concentration	Frequency of Viable Mutants	Relative Percentage							
			Dwarf Mutant	Tall Mutant	Branched Mutant	Early Flowering	Long Pod Mutant	Early Maturing	Late Maturing	High Yielding
Control	Control	–	–	–	–	–	–	–	–	–
Gamma rays	5kR	13.25	2.04	–	3.18	1.17	0.89	2.26	1.32	2.39
	10kR	13.74	1.26	1.05	0.99	4.12	–	0.68	2.46	3.18
	20kR	12.16	–	2.25	0.67	3.20	1.28	0.92	1.36	2.48
EMS (per cent)	0.05	14.71	1.98	2.23	–	3.68	0.98	1.84	2.28	1.72
	0.10	14.96	2.12	–	–	4.19	1.27	2.29	1.68	3.41
	0.15	16.04	2.86	1.92	1.58	2.89	2.12	0.78	1.76	2.13
SA (per cent)	0.01	13.92	0.87	2.98	3.12	2.87	–	1.44	0.96	1.68
	0.02	15.83	3.48	1.67	1.89	0.73	2.24	2.02	2.68	1.12
	0.03	18.34	2.18	1.22	2.24	3.81	1.92	1.62	2.28	2.07

Table 28: Morphological Characters of Promising Viable Mutants in M_2 Generation of *Abelmoschus esculentus* (L.) Moench.

Variety : Parbhani Kranti

Sl.No.	Characters	Control	Dwarf Mutant	Tall Mutant	Branched Mutant	Early Flowering Mutant	Long Pod Mutant	Early Maturing Mutant	Late Maturing Mutant	High Yielding Mutant
1.	Habit	Annual shurb	Annual shurb	Annual shurb	Annual shurb	Annual shurb	Annual shurb	Annual shurb	Annual shurb	Annual shurb
2.	Root	Tap root system	Tap root system	Tap root system	Tap root system	Tap root system	Tap root system	Tap root system	Tap root system	Tap root system
3.	Stem	Singled stemed with sparsely haired	Singled stemed with sparsely haired	Singled stemed with sparsely haired	Branched with sparsely haired	Singled stemed with sparsely haired	Singled stemed with sparsely haired	Singled stemed with sparsely haired	Singled stemed with sparsely haired	Singled stemed with sparsely haired
4.	Number of branches	Nil	Nil	Nil	Four branches	Nil	Nil	Nil	Nil	Nil
5.	Leaf	Deeply lobed with narrow leaflet in top 1/3 portion	Deeply lobed with narrow leaflet	Deeply lobed with narrow leaflet	Deeply lobed with narrow leaflet	Deeply lobed with narrow leaflet	Deeply lobed with narrow leaflet	Deeply lobed with narrow leaflet	Deeply lobed with narrow leaflet	Deeply lobed with narrow leaflet
6.	Inflorescences	Axillary solitary	Axillary solitary	Axillary solitary	Axillary solitary	Axillary solitary	Axillary solitary	Axillary solitary	Axillary solitary	Axillary solitary
7.	Days to first flowering	45	48	42.22	44.84	36.37	48.13	46.18	50.00	49.34
8.	Days to maturity	120	118.04	118	121	116.28	120.28	104.02	130.04	117.68
9.	Plant height (cm)	115	88.72	136.28	110.53	112.26	118.04	115.26	115.00	113.45
10.	Nodes on main stem	07	04	10.67	06.12	7.29	07.13	6.43	7.02	10.33
11.	Flower appearing node	06	04.92	09	5.98	7.34	5.46	5.92	6.85	9.18
12.	Pod length (cm)	12	11.48	12.04	14.08	11.88	20.32	12.03	13.00	13.31

Contd...

Table 28–*Contd...*

Sl.No.	Characters	Control	Dwarf Mutant	Tall Mutant	Branched Mutant	Early Flowering Mutant	Long Pod Mutant	Early Maturing Mutant	Late Maturing Mutant	High Yielding Mutant
13.	Pods per plant	11	10.06	14.89	18.03	12.31	10.33	11.63	10.26	16.43
14.	Seeds per pod	58.42	56.36	62.13	55.24	60.35	62.03	59.00	56.33	57.36
15.	Picking duration	55	52	48.66	49.02	40.63	52.00	51.04	56.17	54.62
16.	Green pod yield	300.00	294.18	326.00	361.29	312.46	306.48	310.00	289.72	448.16
17.	100 seed weight (gm)	6.6	6.3	7.2	6.8	6.10	6.8	5.08	4.20	5.9

5) Long Pod

In these mutants the length of pod increased than control. The length of the pod in these mutants was 20.32 cm as against 17.24 cm in control. They matured at par with control. These mutants showed positive correlation between long pod with green pod yield per plant.

6) Early Maturing Mutant

These mutants matured almost fifteen days earlier than the control population. They matured in 104.02 days quite earlier to control (120 days). There was no significant difference in pod yield per plant (372.63 gm) in such mutants as against (398.27 gm) in control.

7) Late Maturing Mutant

These mutants matured quite late than control. They took 130.04 days for maturity as against 120.00 days in control. They exhibited reduction in 100 seed weight (4.2 gm).

8) High Yielding Mutant

These mutants showed green pod yield higher than control. They showed 448.16 gm of yield per plant against the yield value of 398.27 gm for control. There was a positive correlation between nodes on main stem and pods per plant.

II) Viable Mutants in Variety Arka Anamika (Table 29)

1) Dwarf Mutant

These mutants were characterized by an extreme reduction in plant height. It was 79.43 cm as against 102.26 cm for control. All the characters in this mutant demonstrated reduced values as compared with control.

2) Tall Mutant

The tall mutant showed a height of 128.14 cm as against 102.26 cm in control. These mutants showed positive correlation between height of the plant with nodes on main stem. The period for maturity in this mutant was comparable with that of control.

3) Branched Mutant

These mutants showed 5-6 branches. They demonstrated reduction in number of nodes on branches and reduced height.

4) Early Flowering Mutant

These mutants attained flowering almost eight days earlier than the control population. They flowered in 40.98 days as compared with 49.02 days for control. These mutants showed positive correlation between days to flowering and 100 seed weight (6.5 gm) and days to flowering.

Table 29: Morphological Characters of Promising Viable Mutants in M_2 Generation of *Abelmoschus esculentus* (L.) Moench.

Variety : Arka Anamika

Sl.No.	*Characters*	*Control*	*Dwarf Mutant*	*Tall Mutant*	*Branched Mutant*	*Early Flowering Mutant*	*Long Pod Mutant*	*Early Maturing Mutant*	*Late Maturing Mutant*	*High Yielding Mutant*
1.	Habit	Annual shurb	Annual shurb	Annual shurb	Annual shurb	Annual shurb	Annual shurb	Annual shurb	Annual shurb	Annual shurb
2.	Root	Tap root system	Tap root system	Tap root system	Tap root system	Tap root system	Tap root system	Tap root system	Tap root system	Tap root system
3.	Stem	Splashes of purple pigmentation present on stem with hairs	Purple pigmentation present on stem	Stem with sparasely hair and pigmen-tation	Stem with sparasely hair and pigmen-tation	Stem with pigmen-tation	Purple pigmentation present on stem with hairs	Pigmentation present on stem with sparsly hair	Purple pigmen-tation present on stem with hairs	Pigmentation on stem with sparsly hair
4.	Number of branches	Three to four	Three to four	Three to four	Five to six	Three to four	Three to four	Three to four	Two to three	Five to six
5.	Leaf	Leaves are green, small and deeply lobed and sparsely haired	Green, small and deeply lobed	Green, small and deeply lobed with hair	Green, small and deeply lobed with hair	Green, small and deeply lobed with hair	Leaves are green, small and heart shaped with sparsely haired	Leaves are green and deeply lobed with hair	Leaves are dark green and deeply lobed with hair	Leaves are dark green and heart shaped
6.	Inflorescences	Axillary solitary, five yellow petals	Axillary solitary, five yellow petals	Axillary solitary, five yellow petals	Axillary solitary, five yellow petals	Axillary solitary, five yellow petals	AxillaryA solitary, yellow petals	Axillary solitary, five yellow petals	Axillary solitary, five yellow petals	Axillary solitary, five yellow petals
7.	Days to first flowering	50	48	48.31	51.84	40.98	51	43.44	54.32	50.03
8.	Days to maturity	125	120	122.68	128.00	112.43	116.26	108.28	130.20	120.00
9.	Plant height (cm)	100	79.43	128.14	96.17	102.00	106.02	110.00	102.00	98.17

Contd...

Table 29–*Contd...*

Sl.No.	*Characters*	*Control*	*Dwarf Mutant*	*Tall Mutant*	*Branched Mutant*	*Early Flowering Mutant*	*Long Pod Mutant*	*Early Maturing Mutant*	*Late Maturing Mutant*	*High Yielding Mutant*
10.	Nodes on main stem	8.36	7.98	12.24	6.32	7.02	7.92	7.04	6.92	9.12
11.	Flower appearing node	6.2	5.8	11.63	6.4	5.32	6.84	5.98	6.00	8.75
12.	Pod length (cm)	10.17	10.00	9.87	10.23	11.27	14.32	11.92	10.26	11.02
13.	Pods per plant	9.67	9.02	10.26	14.48	10.00	10.00	8.84	7.92	10.23
14.	Seeds per pod	50.00	45.34	51.59	52.62	48.68	54.82	53.78	46.12	54.36
15.	Picking duration	55.00	54.28	53.66	55.23	45.33	55.33	49.23	60.00	54.83
16.	Green pod yield	300.00	284.67	302.82	312.46	304.61	332.07	292.05	282.76	340.13
17.	100 seed weight (gm)	6.2	6.00	6.4	5.6	6.5	5.8	6.1	6.00	6.3

Figure 9: Dwarf Mutant of Okra (Variety: Parbhani Kranti).

Figure 10: Dwarf Mutant of Okra (Variety: Arka Anamika).

Figure 11: Tall Mutant of Okra (Variety: Parbhani Kranti).

Figure 12: Tall Mutant of Okra (Variety: Arka Anamika).

5) Long Pod

These mutants showed long pod character (14.32 cm) as compared with control (12.18 cm). They matured earlier than that of control. These mutants showed positive correlation between pod length with green pod yield per plant.

6) Early Maturing Mutant

These mutants matured almost ten days earlier than the control population and they matured in 108.28 days and had decreased 100 seed weight (6.1 gm).

7) Late Maturing Mutant

These mutants matured late than control. They took a maturity period of 139.20 days as against 118.42 days for control. They demonstrated reduced pod per plant.

Figure 13: Branched Mutant of Okra (Variety: Parbhani Kranti).

8) High Yielding Mutant

These mutants showed higher pod yield (300.13 gm). These mutants showed positive correlation between nodes and pods on plants.

Quantitative Characters in M_2 and M_3 Generations

In the present programme, eleven quantitative parameters were studied to estimate the induced variability in M_2 and M_3 generations. The different characters studied in Parbhani Kranti and Arka Anamika for both the M_2 and M_3 generations were:

1) Days to flowering
2) Days to maturity
3) Plant height (in cm)
4) Nodes on main stem
5) First flower appearing node
6) Pod length (in cm)
7) Pods per plant
8) Number of seeds per pod

Figure 14: Branched Mutant of Okra (Variety: Arka Anamika).

9) First picking duration
10) Green pod yield (in gm)
11) Hundred Seed weight (in gm).

A thorough statistical analysis was carried out for computing the mean, standard error and coefficient of variation using standard formulae (Mungikar 1977). Moreover the 't' test was used to find out significance. Differences in means between controls and treated populations were estimated to study the amount of variability induced by the three mutagenic treatments.

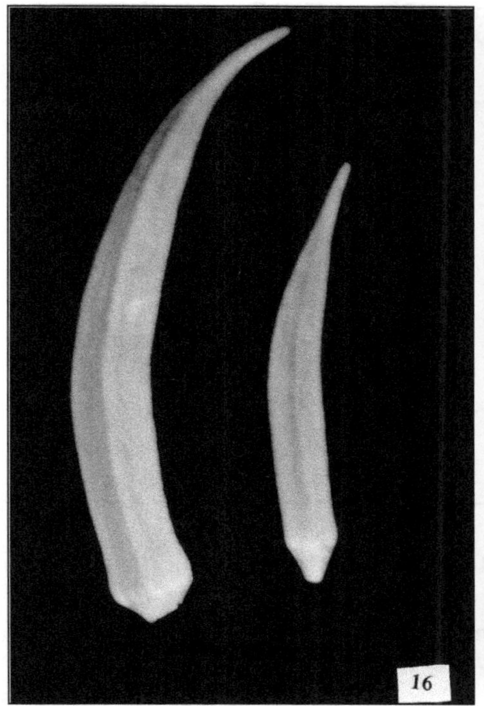

Figure 15: Pod Variability in M_2 Generation of Okra (Variety: Parbhani Kranti).

Figure 16: Pod Variability in M_2 Generation of Okra (Variety: Arka Anamika).

The data pertaining to M_2 and M_3 generations recorded for different polygenic traits in both the varieties of okra are presented separately in tables.

Days to Flowering (Tables 30-33)

The mean values in regard to days to flowering showed early flowering than control, in both the varieties of 0rka, except in gamma rays (10kR and 20kR) of Parbhani Kranti in M_2 generation. Mean values revealed a shift towards negative direction in majority of the treatments having significant variation.

In M_3 generation the mean values shifted in negative direction in Parbhani Kranti and Arka Anamika, respectively. It was early flowering in all treatments of both the varieties of okra. The highest negative shift could be seen in 0.02 per cent SA treatment in Parbhani Kranti and 0.15 per cent EMS treatment in Arka Anamika.

In variety Parbhani Kranti and Arka Anamika majority of the treatments revealed significant negative shift in mean values in M_2 and M_3 generations.

Table 30: The Effect of Mutagens on Days to First Flowering in M_2 Generation in *Abelmoschus esculentus* (L.) Moench.

Variety : Parbhani Kranti

Mutagen	*Dose*	*Mean*	*Shift in Mean*
Control	–	43.00	–
Gamma rays	5kR	42.57	–0.43
	10kR	43.06	+0.06
	20kR	43.14	+0.14
EMS (per cent)	0.05	41.67	–1.33
	0.10	40.98	–2.02
	0.15	40.03	–2.97
SA (per cent)	0.01	42.30	–0.70
	0.02	38.12	–4.88
	0.03	39.74	–3.26

CV = 4.10 per cent ±SE = 0.54

CD at 5 per cent = 1.23 CD at 1 per cent = 1.75

Table 31: The Effect of Mutagens on Days to First Flowering in M_2 Generation in *Abelmoschus esculentus* (L.) Moench.

Variety : Arka Anamika

Mutagen	*Dose*	*Mean*	*Shift in Mean*
Control	–	50.00	–
Gamma rays	5kR	46.56	–3.44
	10kR	47.71	–2.29
	20kR	48.01	–1.99
EMS (per cent)	0.05	45.33	–4.67
	0.10	44.85	–5.15
	0.15	42.79	–7.21
SA (per cent)	0.01	47.00	–3.00
	0.02	49.33	–0.67
	0.03	46.62	–3.38

CV = 4.57 per cent ±SE = 0.68

CD at 5 per cent = 1.53 CD at 1 per cent = 2.21

Days to Maturity (Tables 34-37)

The mean values in regard to days to maturity demonstrated shift towards positive direction in Parbhani Kranti, except 5kR gamma ray, 0.05 per cent EMS and 0.01 per cent and 0.02 per cent SA treatments in M_2 generation. In variety Arka Anamika, majority of mean values demonstrated shift towards negative direction

Table 32: The Effect of Mutagens on Days to First Flowering in M_3 Generation in *Abelmoschus esculentus* (L.) Moench.

Variety : Parbhani Kranti

Mutagen	*Dose*	*Mean*	*Shift in Mean*
Control	–	39.71	–
	5kR	39.21	–0.5
Gamma rays	10kR	38.73	–1.58
	20kR	37.92	–1.79
	0.05	36.98	–2.73
EMS (per cent)	0.10	37.21	–2.5
	0.15	36.43	–3.28
	0.01	38.52	–1.19
SA (per cent)	0.02	30.41	–9.3
	0.03	38.01	–1.7

CV = 7.04 per cent ±SE = 0.84

CD at 5 per cent = 1.89 CD at 1 per cent = 2.73

Table 33: The Effect of Mutagens on Days to First Flowering in M_3 Generation in *Abelmoschus esculentus* (L.) Moench.

Variety : Arka Anamika

Mutagen	*Dose*	*Mean*	*Shift in Mean*
Control	–	46.21	–
	5kR	42.34	–3.87
Gamma rays	10kR	43.62	–2.59
	20kR	42.17	–4.04
	0.05	41.23	–4.98
EMS (per cent)	0.10	40.58	–5.63
	0.15	36.91	–9.3
	0.01	42.71	–3.5
SA (per cent)	0.02	43.42	–2.79
	0.03	44.23	–1.98

CV = 5.85 per cent ±SE = 0.78

CD at 5 per cent = 1.76 CD at 1 per cent = 2.53

except for 0.15 per cent EMS where positive shift in mean value was evident. The majority mean values showing negative shift comprised significant variation in Arka Anamika in M_2 generation.

Table 34: The Effect of Mutagens on Days to Maturity in M_2 Generation in *Abelmoschus esculentus* (L.) Moench.

Variety : Parbhani Kranti

Mutagen	*Dose*	*Mean*	*Shift in Mean*
Control	–	120.07	–
	5kR	119.00	–1.07
Gamma rays	10kR	120.82	+0.75
	20kR	124.02	+3.32
	0.05	118.14	–1.93
EMS (per cent)	0.10	121.26	+0.90
	0.15	130.24	+9.97
	0.01	117.18	–2.89
SA (per cent)	0.02	119.00	–1.07
	0.03	121,00	+0.93

CV = 3.04 per cent ±SE = 1.18

CD at 5 per cent = 2.66 CD at 1 per cent = 3.82

Table 35: The Effect of Mutagens on Days to Maturity in M_2 Generation in *Abelmoschus esculentus* (L.) Moench.

Variety : Arka Anamika

Mutagen	*Dose*	*Mean*	*Shift in Mean*
Control	–	126.24	–
	5kR	125.16	–1.08
Gamma rays	10kR	123.16	–3.08
	20kR	120.33	–5.91
	0.05	108.52	–17.72
EMS (per cent)	0.10	118.23	–8.01
	0.15	126.46	+0.22
	0.01	125.18	–1.06
SA (per cent)	0.02	134.20	+7.96
	0.03	119.15	–7.09

CV = 5.50 per cent ±SE = 2.14

CD at 5 per cent = 4.83 CD at 1 per cent = 6.95

In M_3 generation majority of mean values showed shift towards negative direction, the plants matured earlier than control, in both the varieties of okra, except 0.10 per cent and 0.15 per cent EMS and 0.03 per cent SA in Parbhani Kranti and 0.15 per cent EMS in Arka Anamika.

Table 36: The Effect of Mutagens on Days to Maturity in M_3 Generation in *Abelmoschus esculentus* (L.) Moench.

Variety : Parbhani Kranti

Mutagen	*Dose*	*Mean*	*Shift in Mean*
Control	–	116.70	–
	5kR	112.04	–4.66
Gamma rays	10kR	114.52	–2.18
	20kR	102.47	–14.23
	0.05	113.35	–3.35
EMS (per cent)	0.10	117.62	+0.92
	0.15	129.74	+13.04
	0.01	113.25	–3.45
SA (per cent)	0.02	116.03	–0.67
	0.03	117.52	+0.82

CV = 5.81 per cent ±SE = 2.14

CD at 5 per cent = 4.83 CD at 1 per cent = 6.95

Table 37: The Effect of Mutagens on Days to Maturity in M_3 Generation in *Abelmoschus esculentus* (L.) Moench.

Variety : Arka Anamika

Mutagen	*Dose*	*Mean*	*Shift in Mean*
Control	–	121.72	–
	5kR	120.54	–1.18
Gamma rays	10kR	118.31	–3.41
	20kR	116.23	–5.49
	0.05	100.78	–20.94
EMS (per cent)	0.10	114.50	–7.22
	0.15	122.23	+0.51
	0.01	120.74	–0.99
SA (per cent)	0.02	118.35	–3.37
	0.03	115.57	–6.15

CV = 5.34 per cent ±SE = 1.98

CD at 5 per cent = 4.48 CD at 1 per cent = 6.43

In variety Parbhani Kranti, the gamma ray 20 kR treatment and in Arka Anamika the EMS 0.05 per cent treatment revealed more significant negative shift in mean values in M_3 generation.

Plant Height (Tables 38-41)

The mean value in regard to plant height demonstrated shift towards positive direction at gamma ray and EMS treatments, while it revealed a shift in negative direction at SA treatment in M_2 generation of Parbhani Kranti. In variety Arka Anamika the mean values shifted towards positive direction at EMS and SA treatments, except 0.01 per cent SA which showed a negative shift and in gamma

Table 38: The Effect of Mutagens on Plant Height in M_2 Generation in *Abelmoschus esculentus* (L.) Moench.

Variety : Parbhani Kranti

Mutagen	*Dose*	*Mean*	*Shift in Mean*
Mutagen	Dose	Mean	Shift in Mean
Control	–	110.00	–
	5kR	126.00	+16.00
Gamma rays	10kR	118.17	+8.17
	20kR	113.05	+0.05
	0.05	117.91	+7.91
EMS (per cent)	0.10	120.72	+10.72
	0.15	124.58	+14.58
	0.01	107.16	–2.84
SA (per cent)	0.02	107.34	–2.66
	0.03	108.21	–1.79

CV =6.18 per cent ±SE = 2.26

CD at 5 per cent = 5.10 CD at 1 per cent = 7.34

Table 38: The Effect of Mutagens on Plant Height in M_2 Generation in *Abelmoschus esculentus* (L.) Moench.

Variety : Arka Anamika

Mutagen	*Dose*	*Mean*	*Shift in Mean*
Control	–	102.00	–
	5kR	90.58	–11.42
Gamma rays	10kR	98.62	–3.38
	20kR	101.20	–0.80
	0.05	102.40	–0.40
EMS (per cent)	0.10	104.00	+0.58
	0.15	104.28	+2.28
	0.01	101.02	–9.28
SA (per cent)	0.02	103.47	+1.47
	0.03	105.19	+3.19

CV = 4.16 per cent ±SE = 1.34

CD at 5 per cent = 3.02 CD at 1 per cent = 4.35

Table 40: The Effect of Mutagens on Plant Height in M_3 Generation in *Abelmoschus esculentus* (L.) Moench.

Variety : Parbhani Kranti

Mutagen	*Dose*	*Mean*	*Shift in Mean*
Mutagen	Dose	Mean	Shift in Mean
Control	–	104.00	–
	5kR	129.24	+25.22
Gamma rays	10kR	117.20	+13.2
	20kR	113.68	+9.68
	0.05	114.05	+10.05
EMS (per cent)	0.10	115.91	+11.91
	0.15	118.12	+14.12
	0.01	104.62	+0.62
SA (per cent)	0.02	103.12	–0.88
	0.03	103.36	–0.64

CV = 7.59 per cent ±SE = 2.70

CD at 5 per cent = 6.10 CD at 1 per cent = 8.77

Table 41: The Effect of Mutagens on Plant Height in M_3 Generation in *Abelmoschus esculentus* (L.) Moench.

Variety : Arka Anamika

Control	–	99.21	–
	5kR	93.36	–5.95
Gamma rays	10kR	97.24	–1.97
	20kR	98.27	–0.94
	0.05	96.42	–2.79
EMS (per cent)	0.10	102.17	+2.96
	0.15	104.51	+5.3
	0.01	94.12	–5.09
SA (per cent)	0.02	97.63	–1.58
	0.03	99.45	+0.24

CV = 3.43 per cent ±SE = 1.06

CD at 5 per cent = 2.41 CD at 1 per cent = 3.44

ray treatments, the mean values showed shift towards negative direction in M_2 generation.

In M_3 generation, the variety Parbhani Kranti demonstrated a positive shift in mean values in gamma ray and EMS treatments, while at SA (0.02 per cent and 0.03 per cent) it showed a negative shift in mean values. The significant negatively shifted mean values could be observed at gamma ray 5kR treatment in Arka Anamika, while the SA 0.01 per cent showed significant negative shift in mean values. In both the varieties of okra, the mean values shifted in negative and positive direction at SA treatments.

Nodes on Main Stem (Tables 42-45)

All the treatments employed in the present investigation induced variability in number of nodes on main stem. It indicated the mean values getting shifted towards negative direction than control, except for EMS 0.15 per cent in M_2 generation of

Table 42: The Effect of Mutagens on Number of Nodes on Main Stem in M_2 Generation in *Abelmoschus esculentus* (L.) Moench.

Variety : Parbhani Kranti

Mutagen	*Dose*	*Mean*	*Shift in Mean*
Control	–	11.20	
	5kR	10.68	–0.52
Gamma rays	10kR	10.96	–0.24
	20kR	11.04	–0.16
	0.05	10.36	–0.84
EMS (per cent)	0.10	11.16	–0.04
	0.15	12.24	+1.04
	0.01	10.04	–1.16
SA (per cent)	0.02	9.92	–1.28
	0.03	10.16	–1.04

CV = 6.49 per cent ±SE = 0.24

CD at 5 per cent = 0.54 CD at 1 per cent = 0.78

Table 43: The Effect of Mutagens on Number of Nodes on Main Stem in M_2 Generation in *Abelmoschus esculentus* (L.) Moench.

Variety : Arka Anamika

Mutagen	*Dose*	*Mean*	*Shift in Mean*
Control	–	8.36	
	5kR	8.92	+0.56
Gamma rays	10kR	9.44	+1.08
	20kR	9.68	+1.32
	0.05	9.72	+1.36
EMS (per cent)	0.10	9.86	+1.50
	0.15	10.13	+1.77
	0.01	9.14	+0.78
SA (per cent)	0.02	8.64	+0.28
	0.03	9.84	+1.48

CV = 6.23 per cent ±SE = 0.19

CD at 5 per cent = 0.42 CD at 1 per cent = 0.61

Table 44: The Effect of Mutagens on Number of Nodes on Main Stem in M_3 Generation in *Abelmoschus esculentus* (L.) Moench.

Variety : Parbhani Kranti

Mutagen	*Dose*	*Mean*	*Shift in Mean*
Control	–	6.72	
	5kR	6.76	+0.04
Gamma rays	10kR	6.88	+0.16
	20kR	7.12	+0.04
	0.05	6.17	–0.55
EMS (per cent)	0.10	7.31	+0.59
	0.15	8.24	+1.52
	0.01	6.04	–0.68
SA (per cent)	0.02	4.72	–2.00
	0.03	6.72	

CV = 13.78 per cent ±SE = 0.29

CD at 5 per cent = 0.65 CD at 1 per cent = 0.94

Table 45: The Effect of Mutagens on Number of Nodes on Main Stem in M_3 Generation in *Abelmoschus esculentus* (L.) Moench.

Variety : Arka Anamika

Mutagen	*Dose*	*Mean*	*Shift in Mean*
Control	–	8.02	
	5kR	6.24	–1.78
Gamma rays	10kR	6.72	–1.30
	20kR	8.17	+0.15
	0.05	6.22	–1.80
EMS (per cent)	0.10	6.54	–1.48
	0.15	7.42	–0.60
	0.01	6.48	–1.54
SA (per cent)	0.02	5.26	–2.76
	0.03	6.18	–1.84

CV = 13.37 per cent ±SE = 0.28

CD at 5 per cent = 0.63 CD at 1 per cent = 0.91

Parbhani Kranti, while in Arka Anamika all mean values revealed increased feature over the control. The EMS 0.15 per cent treatment showed significant positive shift in both the varieties of okra.

In M_3 generation the variety Parbhani Kranti revealed the significant positive shift in mean values at EMS 0.15 per cent. The mean values demonstrated positive shift at gamma rays and negative shift at SA treatments. In variety Arka Anamika mean values were reduced than control, except for 20 kR gamma ray treatment. The mean values shifted in negative direction as compared with the control.

First Flower Appearing Node (Tables 46-49)

Both the varieties of okra could be observed with similar positive and negative shifts in mean values in all treatments used in the present investigation. It indicated

Table 46: The Effect of Mutagens on First Flower Appearing Node in M_2 Generation in *Abelmoschus esculentus* (L.) Moench.

Variety : Parbhani Kranti

Mutagen	*Dose*	*Mean*	*Shift in Mean*
Control	–	6.70	
	5kR	6.64	–0.06
Gamma rays	10kR	5.82	–0.88
	20kR	6.13	–0.57
	0.05	6.31	–0.39
EMS (per cent)	0.10	6.85	+0.15
	0.15	7.00	+0.30
	0.01	7.10	+0.40
SA (per cent)	0.02	7.50	+0.80
	0.03	7.90	+1.20

CV = 9.20 per cent ±SE = 0.22

CD at 5 per cent = 0.49 CD at 1 per cent = 0.71

Table 47: The Effect of Mutagens on First Flower Appearing Node in M_2 Generation in *Abelmoschus esculentus* (L.) Moench.

Variety : Arka Anamika

Mutagen	*Dose*	*Mean*	*Shift in Mean*
Control	–	6.20	
	5kR	5.92	–0.32
Gamma rays	10kR	5.84	–0.44
	20kR	6.12	–0.12
	0.05	6.13	–0.13
EMS (per cent)	0.10	6.41	+0.21
	0.15	6.50	+0.30
	0.01	6.32	+0.12
SA (per cent)	0.02	6.50	+0.30
	0.03	6.61	+0.41

CV = 4.11 per cent ±SE = 0.10

CD at 5 per cent = 0.22 CD at 1 per cent = 0.32

Table 48: The Effect of Mutagens on First Flower Appearing Node in M_3 Generation in *Abelmoschus esculentus* (L.) Moench.

Variety : Parbhani Kranti

Mutagen	*Dose*	*Mean*	*Shift in Mean*
Control	–	6.18	
	5kR	5.22	–0.96
Gamma rays	10kR	5.49	–0.69
	20kR	6.03	–0.15
	0.05	5.68	–0.5
EMS (per cent)	0.10	6.12	–0.3
	0.15	6.30	+0.12
	0.01	4.98	–1.2
SA (per cent)	0.02	5.00	–1.18
	0.03	5.36	–0.82

CV = 8.84 per cent ±SE = 0.16

CD at 5 per cent = 0.36 CD at 1 per cent = 0.52

Table 49: The Effect of Mutagens on First Flower Appearing Node in M_3 Generation in *Abelmoschus esculentus* (L.) Moench.

Variety : Arka Anamika

Mutagen	*Dose*	*Mean*	*Shift in Mean*
Control	–	5.26	
	5kR	4.13	–1.13
Gamma rays	10kR	4.86	–0.4
	20kR	5.10	–0.16
	0.05	5.32	+0.6
EMS (per cent)	0.10	5.00	–0.26
	0.15	4.46	–0.8
	0.01	5.00	–0.26
SA (per cent)	0.02	4.92	–0.34
	0.03	4.74	–0.52

CV = 7.42 per cent ±SE = 0.11

CD at 5 per cent = 0.26 CD at 1 per cent = 0.35

mean values getting shifted towards negative direction at gamma ray and 0.05 per cent EMS treatments and mean values shifting towards positive direction at 0.10 per cent, 0.15 per cent EMS and 0.01 per cent, 0.02 per cent and 0.03 per cent SA treatments in variety Parbhani Kranti and Arka Anamika, respectively in M_2 generation. The negative shift in mean values at 10kR gamma rays showed significance in both the varieties of okra.

In M_3 generation the mean values got reduced than control, except 0.15 per cent EMS in Parbhani Kranti and 0.05 per cent EMS in Arka Anamika varieties of okra. In Parbhani Kranti the significant mean values could be observed at 5kR, 10kR gamma rays, 0.05 per cent EMS and 0.01 per cent, 0.02 per cent and 0.03 per cent SA treatments, while in Arka Anamika except 20kR gamma ray and 0.05 per cent EMS treatment at all the remaining treatments significant negative mean values could be clearly seen.

Pod Length (Tables 50-53)

The mean values in regard to pod length demonstrated positive shift in mean values at 0.10 per cent and 0.15 per cent EMS treatments in Parbhani Kranti, while at remaining treatments the negative shift in mean values could be seen as compared with control. In Arka Anamika variety the mean values reduced at EMS and SA treatments and increased at gamma ray treatment in M_2 generation. The gamma rays 5kR and 10kR showed significant positive shift in mean values in Arka Anamika, while 0.10 per cent and 0.15 per cent of EMS showed significant positive shift in mean values in variety Parbhani Kranti. The highest positive shift in mean values could be observed at 0.15 per cent EMS and 5kR gamma rays in variety Parbhani Kranti and Arka Anamika, respectively.

The majority of the reduced mean values could be observed at M_3 generation. In variety Parbhani Kranti, the mean values showed significant positive shift at 0.10 per cent and 0.15 per cent EMS treatments. While in Arka Anamika at 5kR gamma ray treatment significant positive shift in mean values could be observed. The reduced mean values could be seen at gamma ray and SA treatments in Parbhani Kranti and in variety Arka Anamika all the EMS concentrations showed reduced mean values than the control.

Table 50: The Effect of Mutagens on Pod Length in M_2 Generation in *Abelmoschus esculentus* (L.) Moench.

Variety : Parbhani Kranti

Mutagen	*Dose*	*Mean*	*Shift in Mean*
Control	–	17.2	
	5kR	17.1	–0.01
Gamma rays	10kR	16.2	–1.00
	20kR	15.9	–1.30
	0.05	16.8	–0.40
EMS (per cent)	0.10	18.3	+1.1
	0.15	18.4	+1.2
	0.01	15.8	–1.4
SA (per cent)	0.02	16.1	–1.1
	0.03	16.00	–1.2

CV = 5.44 per cent ±SE = 0.32

CD at 5 per cent = 0.72 CD at 1 per cent = 1.04

Table 51: The Effect of Mutagens on Pod Length in M_2 Generation in *Abelmoschus esculentus* (L.) Moench.

Variety : Arka Anamika

Mutagen	*Dose*	*Mean*	*Shift in Mean*
Control	–	12.17	
	5kR	14.34	+2.17
Gamma rays	10kR	13.88	+1.71
	20kR	12.44	+0.27
	0.05	10.40	–1.77
EMS (per cent)	0.10	10.90	–1.27
	0.15	11.80	–0.37
	0.01	9.70	–2.47
SA (per cent)	0.02	10.00	–2.17
	0.03	10.20	–1.97

CV = 14.04 per cent ±SE = 0.52

CD at 5 per cent = 1.17 CD at 1 per cent = 1.69

Table 52: The Effect of Mutagens on Pod Length in M_3 Generation in *Abelmoschus esculentus* (L.) Moench.

Variety : Parbhani Kranti

Mutagen	*Dose*	*Mean*	*Shift in Mean*
Control	–	15.4	
	5kR	15.1	–0.3
Gamma rays	10kR	14.7	–0.7
	20kR	14.5	–0.9
	0.05	15.3	–0.1
EMS (per cent)	0.10	16.5	+1.1
	0.15	19.2	+3.8
	0.01	14.4	–1.0
SA (per cent)	0.02	13.3	–2.1
	0.03	13.5	–1.9

CV = 11.10 per cent ±SE = 0.53

CD at 5 per cent = 1.21 CD at 1 per cent = 1.72

Pods per Plant (Tables 54-57)

The mean values in regard to number of pods per plant in M_2 generation showed positive shift at 0.10 per cent and 0.15 per cent EMS treatments in variety Parbhani Kranti, while at remaining treatments a negative shift in mean values could be noted. In variety Arka Anamika except 5kR, 20kR gamma ray and 0.01 per cent

Table 53: The Effect of Mutagens on Pod Length in M_3 Generation in *Abelmoschus esculentus* (L.) Moench.

Variety : Arka Anamika

Mutagen	*Dose*	*Mean*	*Shift in Mean*
Control	–	9.15	
	5kR	11.12	+1.97
Gamma rays	10kR	9.47	+0.32
	20kR	9.12	–0.03
	0.05	7.42	–1.73
EMS (per cent)	0.10	8.15	–1.00
	0.15	8.48	–0.67
	0.01	9.31	+0.16
SA (per cent)	0.02	8.68	–0.47
	0.03	8.20	–0.95

CV = 11.22 per cent ±SE = 0.32

CD at 5 per cent = 0.72 CD at 1 per cent = 1.04

SA treatments, at the remaining treatments a positive shift in mean values could be observed. The EMS treatments at 0.15 per cent and 0.05 per cent, 0.10 per cent, 0.15 per cent showed significant positive shift in mean values in variety Parbhani Kranti and Arka Anamika, respectively. The highest positive mean values could be observed at 0.15 per cent EMS in variety Arka Anamika.

Table 54: The Effect of Mutagens on Number of Pods per Plant in M_2 Generation in *Abelmoschus esculentus* (L.) Moench.

Variety : Parbhani Kranti

Mutagen	*Dose*	*Mean*	*Shift in Mean*
Control	–	16.00	
	5kR	14.40	–1.6
Gamma rays	10kR	12.64	–3.36
	20kR	13.82	–2.18
	0.05	15.62	–0.38
EMS (per cent)	0.10	16.03	+0.03
	0.15	16.92	+0.92
	0.01	12.98	–3.02
SA (per cent)	0.02	13.21	–2.79
	0.03	13.96	–2.04

CV = 10.21 per cent ±SE = 0.48

CD at 5 per cent = 1.08 CD at 1 per cent = 1.56

Table 55: The Effect of Mutagens on Number of Pods per Plant in M_2 Generation in *Abelmoschus esculentus* (L.) Moench.

Variety : Arka Anamika

Mutagen	*Dose*	*Mean*	*Shift in Mean*
Control	–	9.67	
	5kR	8.57	–1.1
Gamma rays	10kR	10.2	+0.53
	20kR	9.4	–0.27
	0.05	10.6	+0.93
EMS (per cent)	0.10	12.08	+2.41
	0.15	12.14	+2.47
	0.01	9.2	–0.47
SA (per cent)	0.02	9.8	+0.13
	0.03	10.3	+0.63

CV = 11.42 per cent ±SE = 0.39

CD at 5 per cent = 0.88 CD at 1 per cent = 1.26

Table 56: The Effect of Mutagens on Number of Pods per Plant in M_3 Generation in *Abelmoschus esculentus* (L.) Moench.

Variety : Parbhani Kranti

Mutagen	*Dose*	*Mean*	*Shift in Mean*
Control	–	11.51	
	5kR	11.42	–0.09
Gamma rays	10kR	9.17	–2.34
	20kR	9.87	–1.64
	0.05	11.24	-0.27
EMS (per cent)	0.10	12.17	–0.66
	0.15	12.22	+0.71
	0.01	9.82	–1.69
SA (per cent)	0.02	10.47	–1.04
	0.03	10.97	–0.54

CV = 9.45 per cent ±SE = 0.33

CD at 5 per cent = 0.75 CD at 1 per cent = 1.08

In M_3 generation the mean values decreased at gamma ray and SA and 0.05 per cent EMS treatments in variety Parbhani Kranti, whereas at 0.10 per cent, 0.15 per cent EMS treatments the significant positive shift in mean values could be observed. In variety Arka Anamika the significant positive shift in mean values could be seen at 10kR gamma rays, 0.15 per cent EMS and 0.02 per cent, 0.03 per cent SA treatments. The highest positive mean value in Arka Anamika was observable at 0.15 per cent EMS and 0.03 per cent SA treatments.

Table 57: The Effect of Mutagens on Number of Pods per Plant in M_3 Generation in *Abelmoschus esculentus* (L.) Moench.

Variety : Arka Anamika

Mutagen	*Dose*	*Mean*	*Shift in Mean*
Control	–	6.74	
	5kR	6.12	–0.62
Gamma rays	10kR	8.72	+1.98
	20kR	6.40	–0.34
	0.05	6.17	–0.57
EMS (per cent)	0.10	6.31	–0.43
	0.15	8.47	+1.73
	0.01	6.25	–0.49
SA (per cent)	0.02	7.39	+0.65
	0.03	8.13	+1.39

CV = 14.47 per cent ±SE = 0.35

CD at 5 per cent = 0.80 CD at 1 per cent = 1.14

Seeds per Pod (Tables 58-61)

The mean values in regard to number of seeds per pod in M_2 generation of variety Parbhani Kranti showed negative shift in all treatments, except 0.15 per cent EMS. While in variety Arka Anamika the mean values showed positive shift in mean values in all the treatments except 0.02 per cent SA.

Table 58: The Effect of Mutagens on Seeds per Pod in M_2 Generation in *Abelmoschus esculentus* (L.) Moench.

Variety : Parbhani Kranti

Mutagen	*Dose*	*Mean*	*Shift in Mean*
Control	–	63.00	
	5kR	52.68	–10.32
Gamma rays	10kR	52.48	–10.52
	20kR	50.40	–12.6
	0.05	53.36	–9.64
EMS (per cent)	0.10	59.24	–3.76
	0.15	64.44	+1.44
	0.01	50.00	–13.00
SA (per cent)	0.02	48.00	–15.00
	0.03	57.00	–6.00

CV = 10.21 per cent ±SE = 1.78

CD at 5 per cent = 4.00 CD at 1 per cent = 5.78

Table 59: The Effect of Mutagens on Seeds per Pod in M_2 Generation in *Abelmoschus esculentus* (L.) Moench.

Variety : Arka Anamika

Mutagen	*Dose*	*Mean*	*Shift in Mean*
Control	–	50.00	
	5kR	52.62	+2.62
Gamma rays	10kR	52.33	+2.33
	20kR	51.51	+1.51
	0.05	53.12	+3.12
EMS (per cent)	0.10	56.37	+6.37
	0.15	61.52	+11.52
	0.01	52.43	+2.43
SA (per cent)	0.02	49.28	–0.72
	0.03	50.89	+0.89

CV = 6.72 per cent ±SE = 1.15

CD at 5 per cent = 2.61 CD at 1 per cent = 3.73

Table 60: The Effect of Mutagens on Seeds per Pod in M_3 Generation in *Abelmoschus esculentus* (L.) Moench.

Variety : Parbhani Kranti

Mutagen	*Dose*	*Mean*	*Shift in Mean*
Control	–	56.71	
	5kR	47.45	–9.26
Gamma rays	10kR	48.74	–7.97
	20kR	49.53	–7.18
	0.05	47.56	–9.15
EMS (per cent)	0.10	53.42	–3.29
	0.15	59.74	+3.03
	0.01	44.72	–11.99
SA (per cent)	0.02	43.12	–13.59
	0.03	53.64	–3.07

CV = 10.49 per cent ±SE = 1.68

CD at 5 per cent = 3.80 CD at 1 per cent = 5.46

In M_2 and M_3 generations the significant positive shift in mean values could be observed at 0.15 per cent EMS in variety Parbhani Kranti, while in Arka Anamika the significant positive shift in mean values could be recorded at gamma ray, EMS and 0.01 per cent SA treatments in M_2 generation, and 20kR gamma ray, 0.10 per cent and 0.15 per cent EMS treatments in M_3 generation. The highest positive shift in mean values was noticeable at 0.15 per cent EMS in Parbhani Kranti, in M_2 and M_3 generations.

Table 61: The Effect of Mutagens on Seeds per Pod in M_2 Generation in *Abelmoschus esculentus* (L.) Moench.

Variety : Arka Anamika

Mutagen	*Dose*	*Mean*	*Shift in Mean*
Control	–	45.24	
	5kR	46.71	+1.47
Gamma rays	10kR	47.53	+2.29
	20kR	48.43	+3.19
	0.05	47.53	+2.29
EMS (per cent)	0.10	52.71	+7.47
	0.15	55.74	+10.5
	0.01	47.35	+2.11
SA (per cent)	0.02	43.84	–1.4
	0.03	45.92	+0.68

CV = 7.39 per cent ±SE = 1.14

CD at 5 per cent = 2.58 CD at 1 per cent = 3.70

First Picking Duration (Tables 62-65)

The mean value in regard to days to first picking in M_2 generation showed reduction in values than control in both the varieties of okra, except for 0.02 per cent Sa treatment in Arka Anamika. The variety Parbhani Kranti showed significant

Table 62: The Effect of Mutagens on Days of First Picking in M_2 Generation in *Abelmoschus esculentus* (L.) Moench.

Variety : Parbhani Kranti

Mutagen	*Dose*	*Mean*	*Shift in Mean*
Control	–	53.00	
	5kR	47.30	–5.7
Gamma rays	10kR	48.40	–4.6
	20kR	48.44	–4.56
	0.05	45.24	–7.76
EMS (per cent)	0.10	44.16	–8.84
	0.15	44.10	–8.9
	0.01	49.26	–3.74
SA (per cent)	0.02	44.16	–8.84
	0.03	46.32	–6.68

CV = 6.05 per cent ±SE = 0.91

CD at 5 per cent = 2.05 CD at 1 per cent = 2.95

Table 63: The Effect of Mutagens on Days of First Picking in M_2 Generation in *Abelmoschus esculentus* (L.) Moench.

Variety : Arka Anamika

Mutagen	*Dose*	*Mean*	*Shift in Mean*
Control	–	5.00	
	5kR	52.38	–2.62
Gamma rays	10kR	53.43	–2.57
	20kR	54.56	–0.44
	0.05	51.04	–3.96
EMS (per cent)	0.10	50.63	–4.37
	0.15	48.38	–6.62
	0.01	53.24	–1.76
SA (per cent)	0.02	55.33	+0.33
	0.03	52.38	–2.62

CV = 4.12 per cent ±SE = 0.75

CD at 5 per cent = 1.69 CD at 1 per cent = 2.43

Table 64: The Effect of Mutagens on Days of First Picking in M_3 Generation in *Abelmoschus esculentus* (L.) Moench.

Variety : Parbhani Kranti

Mutagen	*Dose*	*Mean*	*Shift in Mean*
Control	–	46.00	
	5kR	42.57	–3.43
Gamma rays	10kR	42.73	–3.27
	20kR	43.45	–2.55
	0.05	40.42	–5.58
EMS (per cent)	0.10	38.12	–7.88
	0.15	37.85	–8.15
	0.01	41.48	–4.52
SA (per cent)	0.02	38.42	–7.58
	0.03	36.92	–9.08

CV = 7.24 per cent ±SE = 0.95

CD at 5 per cent = 1.92 CD at 1 per cent = 2.76

negative shift in mean values for picking duration in all the treatments. In variety Arka Anamika the 5kR and 10kR gamma rays, 0.05 per cent, 0.10 per cent and 0.15 per cent EMS revealed should significant negative shift in mean values.

In M_3 generation, all the treatments in both the varieties of okra showed reduced mean values than control. Majority of them were having significant mean values for first picking duration.

Table 65: The Effect of Mutagens on Days of First Picking in M_3 Generation in *Abelmoschus esculentus* (L.) Moench.

Variety : Arka Anamika

Mutagen	*Dose*	*Mean*	*Shift in Mean*
Control	–	51.72	
	5kR	47.52	–4.2
Gamma rays	10kR	48.31	–3.41
	20kR	49.26	–2.46
	0.05	47.92	–3.8
EMS (per cent)	0.10	45.34	–6.38
	0.15	44.21	–7.51
	0.01	48.41	–3.31
SA (per cent)	0.02	50.23	–1.49
	0.03	49.49	–2.25

CV = 4.58 per cent ±SE = 0.70

CD at 5 per cent = 1.58 CD at 1 per cent = 2.27

Green Pod Yield (Tables 66-69)

The mean values for green pod yield showed increased feature in majority of treatments in both the varieties of okra, in M_2 generation. In variety Parbhani Kranti, the mean values exhibited shift in positive direction in all the treatments. The mean values demonstrated a positive shift than control in variety Arka Anamika except 0.02 per cent SA treatment.

Table 66: The Effect of Mutagens on Green Pod Yield per Plant in M_2 Generation in *Abelmoschus esculentus* (L.) Moench.

Variety : Parbhani Kranti

Mutagen	*Dose*	*Mean*	*Shift in Mean*
Control	–	400.00	
	5kR	349.13	–50.87
Gamma rays	10kR	368.63	–31.37
	20kR	401.52	+1.02
	0.05	421.14	+21.14
EMS (per cent)	0.10	439.16	+39.16
	0.15	448.00	+48.00
	0.01	356.22	–43.78
SA (per cent)	0.02	308.43	–91.97
	0.03	379.72	–20.28

CV = 11.20 per cent ±SE = 13.72

CD at 5 per cent = 31.00 CD at 1 per cent = 44.5

Table 67: The Effect of Mutagens on Green Pod Yield per Plant in M_2 Generation in *Abelmoschus esculentus* (L.) Moench.

Variety : Arka Anamika

Mutagen	*Dose*	*Mean*	*Shift in Mean*
Control	–	300.00	
	5kR	289.04	–10.96
Gamma rays	10kR	250.57	–49.43
	20kR	242.68	–57.32
	0.05	262.42	–37.58
EMS (per cent)	0.10	278.22	–21.78
	0.15	302.31	+2.31
	0.01	270.26	–29.74
SA (per cent)	0.02	290.00	–10.00
	0.03	272.12	–27.88

CV = 10.3 per cent ±SE = 8.66

CD at 5 per cent = 20.02 CD at 1 per cent = 28.14

Table 68: The Effect of Mutagens on Green Pod Yield per Plant in M_3 Generation in *Abelmoschus esculentus* (L.) Moench.

Variety : Parbhani Kranti

Mutagen	*Dose*	*Mean*	*Shift in Mean*
Control	–	390.24	
	5kR	342.53	–47.71
Gamma rays	10kR	362.41	–28.83
	20kR	394.51	+4.27
	0.05	417.55	+27.31
EMS (per cent)	0.10	434.22	+43.98
	0.15	442.17	+51.93
	0.01	352.11	–38.13
SA (per cent)	0.02	319.42	–70.82
	0.03	374.73	–15.51

CV = 10.52 per cent ±SE = 12.75

CD at 5 per cent = 28.81 CD at 1 per cent = 41.43

In M_3 generation the significant positive shift in mean values was observable at 0.05 per cent, 0.10 per cent and 0.15 per cent EMS treatments in variety Parbhani Kranti and at 0.15 per cent EMS and 0.03 per cent SA treatment in variety Arka Anamika. At remaining treatments the mean values with positive and negative shifting feature could be evidently seen.

Table 69: The Effect of Mutagens on Green Pod Yield per Plant in M_3 Generation in *Abelmoschus esculentus* (L.) Moench.

Variety : Arka Anamika

Mutagen	*Dose*	*Mean*	*Shift in Mean*
Control	–	280.15	
	5kR	282.52	+2.37
Gamma rays	10kR	244.72	–35.43
	20kR	240.15	–40.00
	0.05	254.13	–26.02
EMS (per cent)	0.10	273.12	–7.03
	0.15	294.15	+14.00
	0.01	262.24	–17.91
SA (per cent)	0.02	284.12	+3.97
	0.03	295.53	+15.38

CV = 7.19 per cent ±SE = 6.27

CD at 5 per cent = 14.17 CD at 1 per cent = 20.37

Hundred Seed Weight (Tables 70-73)

The mean values in regard to 100 seed weight showed negative shift at EMS (all concentrations), 10kR and 20kR gamma rays and 0.01 per cent and 0.03 per cent SA treatments. In variety Arka Anamika at 0.01 per cent, 0.02 per cent and 0.03 per cent SA and 0.10 per cent, 0.15 per cent EMS treatments, positive shift in mean values could be seen, while at gamma rays and 0.05 per cent EMS treatments, negative shift in mean values in M_2 generation have been noticed.

Table 70: The Effect of Mutagens on 100 Seed Weight in M_2 Generation in *Abelmoschus esculentus* (L.) Moench.

Variety : Parbhani Kranti

Mutagen	*Dose*	*Mean*	*Shift in Mean*
Control	–	6.6	
	5kR	6.7	+0.1
Gamma rays	10kR	6.5	–0.1
	20kR	6.0	–0.6
	0.05	6.2	–0.4
EMS (per cent)	0.10	5.8	–0.8
	0.15	4.2	–2.4
	0.01	6.5	–0.1
SA (per cent)	0.02	6.9	+0.3
	0.03	6.2	–0.4

CV = 12.41 per cent ±SE = 0.25

CD at 5 per cent = 0.56 CD at 1 per cent = 0.81

Table 71: The Effect of Mutagens on 100 Seed Weight in M_2 Generation in *Abelmoschus esculentus* (L.) Moench.

Variety : Arka Anamika

Mutagen	*Dose*	*Mean*	*Shift in Mean*
Control	–	6.2	
	5kR	5.9	–0.3
Gamma rays	10kR	5.8	–0.4
	20kR	5.1	–1.1
	0.05	6.1	–0.1
EMS (per cent)	0.10	6.4	+0.2
	0.15	6.5	+0.3
	0.01	6.3	+0.1
SA (per cent)	0.02	6.5	+0.3
	0.03	6.6	+0.4

CV = 7.33 per cent ±SE = 0.15

CD at 5 per cent = 0.32 CD at 1 per cent = 0.48

Table 72: The Effect of Mutagens on 100 Seed Weight in M_3 Generation in *Abelmoschus esculentus* (L.) Moench.

Variety : Parbhani Kranti

Mutagen	*Dose*	*Mean*	*Shift in Mean*
Control	–	5.2	
	5kR	5.3	+0.1
Gamma rays	10kR	5.4	+0.2
	20kR	4.9	–0.3
	0.05	5.7	+0.5
EMS (per cent)	0.10	3.8	–1.4
	0.15	3.4	–1.8
	0.01	5.1	–0.1
SA (per cent)	0.02	5.8	+0.6
	0.03	5.9	+0.7

CV = 16.46 per cent ±SE = 0.26

CD at 5 per cent = 0.58 CD at 1 per cent = 0.84

In variety Parbhani Kranti, all the treatments showed positive shift in mean values except 20kR gamma rays, 0.10 per cent and 0.15 per cent EMS, 0.01 per cent SA treatments in M_3 generation, while in Arka Anamika negative shift in mean values could be noted except for 20kR gamma rays and 0.15 per cent EMS treatments. It showed significant negative shift in mean values at 0.10 per cent and 0.15 per cent

EMS in variety Parbhani Kranti, and 0.01 per cent and 0.02 per cent SA, 0.05 per cent EMS, and 5kR and 10kR gamma rays in variety Arka Anamika, respectively.

Table 73: The Effect of Mutagens on 100 Seed Weight in M_3 Generation in *Abelmoschus esculentus* (L.) Moench.

Variety : Arka Anamika

Mutagen	*Dose*	*Mean*	*Shift in Mean*
Control	–	5.3	
	5kR	4.7	–0.6
Gamma rays	10kR	4.3	–1.00
	20kR	5.4	+0.1
	0.05	4.8	–0.5
EMS (per cent)	0.10	5.2	–0.1
	0.15	5.7	+0.4
	0.01	4.4	–0.9
SA (per cent)	0.02	4.5	–0.8
	0.03	4.9	–0.4

CV = 9.47 per cent ±SE = 0.15

CD at 5 per cent = 0.35 CD at 1 per cent = 0.48

Chapter 5

Discussion on Findings

Studies in M_1 Generation

1. Seed Germination

In the present investigation, the germination percentage revealed reduction in gamma ray treatment in both the varieties *viz.* Parbhani Kranti and Arka Anamika of okra. The EMS and SA treatments also showed reduction in germination percentage in case of variety Parbhani Kranti. Whereas the same treatments stimulated germination percentage in variety Arka Anamika. This showed a distinct varietal variation in regard to the mutagenic sensitivity.

The decrease in germination percentage with an increase in gamma ray dose has been reported earlier in okra Heringa (1995) reported reduction in germination with X-ray irradiated pea crop.

The gamma rays retarded the rate of germination and percentage of germination also tended to decrease with an increase in dose (Nandpuri *et al.,* 1971). Sax and Swanson (1941) stated that induced chromosomal alterations are the chief cause of injury in individual cells, but in reaction of tissues to the radiations, the physiological effects may also play an important role. The sensitivity of tomato seeds to X-rays increased tenfold as the water content in seeds decreased (Gladstone *et al.,* 1959).

Cardona *et al.* (1960) recorded that there was reduction in germination percentage in case of X-ray treated seeds of kidney bean.The pertinent researchers opined that this might have been caused by the inhibitions of developmental processes as a result of increase in the intensity of irradiation.

The rate of germination of treated seeds was retarted as compared with that of untreated seeds because the gamma rays disturbed the balance of metabolic processes [Jonson 1928].

An increase in germination percentage after gamma irradiation has been recorded by Dutta (1969), Rao (1983) in okra and Vishnu Swaroop and Gill (1968) in French bean.

Several researchers have reviewed the effects of alkylating agents and their mechanism of action in biological systems. They include, Ross (1962), Loveless (1966), Fishbein *et al.* (1970), Lawley (1973 and 1974) and Sun and Singer (1975). The decrease in germination percentage has been ascribed mainly to the lethality generated in the seeds due to physiological injuries, chromosomal aberrations and the effect of hydrolytic products of the mutagen.

Aman (1968) explained that the endogenous growth regulators play an important role in seed germination and there exists a striking balance between the promoters and the inhibitors. Meherchandani (1975) attributed the reduction in germination to the disturbance of this balance.

Brock (1965) and Larik (1975) have pinpointed that the mutagen induced gross chromosomal breakages may be affecting the germination percentage. Selim *et al.* (1974) have attributed reduction in germination with an increase in radiation exposure to the enhanced production of active radicals which are concerned with lethality. According to Gaul (1964) the pertinent effect is produced mainly through physiological damage.

In the present investigation it was found that the two varieties of okra responded differentially to EMS, SA and gamma ray treatments in respect of seed germination.

The intervarietal variation as regards the radiosensitivity has been reported by Gelin *et al.* (1958) in Pea, Ashri and Herzog (1972) in Pea nut, Kothekar (1987) in *Coriandum sativum*, Kothekar (1989) in moth bean, Hakande (1992) in winged bean and Satpute (1994) in safflower.

According to Ashri and Herzog (1972) the differential varietal radiosensitivity originates from the sites of metabolic processes affected at the embryonic level. Raghuvanshi and Singh (1979) recorded an inverse correlation between dose and germination in all the varieties of *Impatients balsamina* except the purple variety. Based on this it was inferred that a variation of physiological condition within one and same variety might affect the radiosensitivity.

In the present investigation the stimulation of germination percentage at 0.10 per cent EMS and 0.01 per cent, 0.03 per cent SA in variety Arka Anamika could be due to less damage of cell material and the resultant differential chemosensitivity feature.

2. Survival of Plants at Maturity

The extent of survival of plants is considered as one of the most reliable indices in evaluating the effect of any mutagen. The data on plant survival in the present investigation indicated inhibitory effect at gamma ray treatment in both the varieties of okra. The lowest survival values were induced by 20kR gamma ray dose in case of variety Parbhani Kranti and Arka Anamika of okra. It indicated the stimulated values in majority of the EMS and SA treatments in both the varieties of okra.

Malani *et al.* (1933) recorded higher dose of gamma rays resulting in reduced survival of plants in M_2 generation. Similar observations have been recorded by Majid (1975) in tomato with gamma rays and chemical treatment, and Thakur *et al.* (1980) in chilli with X-rays.

The increased lethality values at higher concentrations of different mutagens have been attributed to the injuries generated by physiological imbalance/ chromosomal aberrations (Nilan *et al.*, 1964; Scossirolli *et al.*,.1966, Blixt 1967, and Narayana and Konzak 1969) and due to genetic damage (Davis and Evans 1966). Gaul (1964) stated that positive correlation between increasing mutagen dose and M1 plant survival may be due to alteration at the physiological and cytological levels leading to chromosomal and extrachromosomal injuries.

Several other researchers like Heringa (1964) in Pea; Zannone (1965) in *Vicia*, Sree Ramulu (1971) *in Sorghum*, Kaul and Bhan (1971) in rice, More (1992) in *Medicago sativa* and Satpute (1994) in *Carthamus tinctorius* reported reduced survival after mutagenic treatments, resulting in higher rates of lethality.

3. Leaf Morphological Changes

Leaf morphological changes could be seen in the M1 population of both Parbhani Kranti and Arka Anamika okra varieties. The pertinent variations comprised reduction in size of leaves and suppression of leaf lamina. The frequency of leaf morphological changes carrying plants showed correlation with increasing concentration of mutagen. The frequency of leaf morphological changes carrying plants increased in gamma ray treatment in both the varieties of okra. The 20kR gamma ray treatment showed maximum frequency of leaf changes in Parbhani Kranti and Arka Anamika okra varieties.

Bhasha *et al.* (1988) reported the morphological changes causing notching of leaf tip by gamma ray and sodium azide treatment. Jambhale and Nerkar (1980) reported compound leaves with 3-5 leaflets by gamma irradiation, having leaflets which are narrow and serrated. Venkatramani (1962) and Kalia and Padda (1962) reported inheritance of leaf lobation.

The changes in leaf shape and size have been reported by a number of investigators *e.g.*, in beans (Genter and Brown, 1942), Pea (Gelin 1954), Soybean (Zacharias, 1956), black gram (Jana, 1962 and Appa Rao and Jana 1976), winged bean (Hakande 1992) leucern (More 1992), Kothekar (1978) in *Solanum nigrum*, Deshpande (1980) in *Momordica charantia* and Satpute (1994) in safflower.

The changes in morphology of leaflets induced by mutagens could be due to changes in physiological and metabolic activities of the developing primordia and the consequent alterations in leaf morphology. Joshua *et al.* (1972) have correlated development of leaf abnormalities to the pleiotropic action of mutated genes. Chaturvedi and Singh (1978) found bifoliate, tetrafoliate, pentafoliate and wrinkled leaves in *Phaseolus aureus* after EMS and DES treatments.

In the present study the leaf morphological changes may have developed due to physiological disturbances and developmental alterations produced in both the varieties of okra.

4 Pollen Sterility

The pollen sterility increased in varieties Parbhani Kranti and Arka Anamika with increasing concentrations of EMS and SA treatments. In case of gamma ray treatment in Arka Anamika, pollen sterility decreased with increasing concentration.

Pollen sterility which is an index of disruption in reproductive mechanism increased with an increase in dose of gamma rays in chilli (Ramana Rao *et al.*, 1991). Nandpuri *et al.* (1971) reported increased pollen sterility with an increase in dose of mutagen in okra. The gamma rays produced more effect than EMS in terms of pollen sterility (Jehangir and Chandrasekaran, 1978).

The increase in pollen sterility by ionizing radiations has been reported by Lindstrom (1933), Smith (1938), Morris (1952), Patil and Bora (1961), Hermalin (1967) Chang and Hsich (1957) and Kothekar and Dnyansagar (1985). The induced pollen sterility by chemical mutagens also has been noted by Froese Gertzen *et al.* (1964), Gaul *et al.* (1966), Sato and Gaul (1967), Gohal *et al.* (1972), Kothekar (1978), Deshpande (1980), Hakande (1992), More (1992) and Satpute (1994) in different plant systems.

According to Kivi (1962) the sterility in M_1 generation is indicative of the genetical effectiveness of mutagenic treatment. Sterility counts are considered to be better for calculating the sum of chromosomal aberrations than merely studying the meiotic chromosomes (Gaul, 1977). The major cryptic changes in meiosis due to mutagenic treatments have been implicated for pollen sterility (Wanjari and Kutarekar, 1977).

Sato and Gaul (1967) have divided the pollen sterility induced by EMS in three categories (1) Chromosomal (2) Genic and (3) Purely physiological.

Singh and Chaudhari (1972) observed increased pollen sterility in cluster bean and stated that poor growth, chlorophyll deficiencies and chromosomal abnormalities could be the probable reasons for the pertinent development.

Sudhakaran (1971) has concluded that pollen sterility might represent the cumulative result of various aberrant, meiotic stages as well as physiological and genetic damage induced by the breakage of chromosome through fragmentation caused by the antimetabolic agents in the cell.

5. Chlorophyll Deficient Sectors

The M_1 population of okra revealed the chlorophyll deficient sectors in majority of mutagenic treatments. The frequency of sector carrying plants was highest in 0.03 per cent SA treatment in both the varieties of okra *viz.* Parbhani Kranti and Arka Anamika. The 0. 10 per cent EMS treatment showed least number of sectorial plants in variety Parbhani Kranti while 10kR, 20kR gamma ray and 0.02 per cent SA treatments showed the same feature in variety Arka Anamika.

Jehangir and Chandrasekaran (1978) reported that the chlorophyll chimeras in M_1 increased with an increase in the dose of mutagen.

The mutagen induced chlorophyll deficient sectors embody one of the prime biological effects. This effect however is restricted to M_1 generation only. Gaul (1958) and Goud (1967) opined that a differential response of embryonic cells to mutagen causes chimerism. The development of a chlorophyll chimeric plant becomes possible when a sector of the multicellular embryo becomes mutated.

In the present investigation, the chlorophyll chimeric plants did not show true breeding nature in M_2 generation. Such non-heritable chimeras have been reported by Mackey (1954), Sjodin (1962), Swaminathan (1963), Ramanna and Natarajan (1965), Deshpande (1980), Hakande (1992) and More (1992).

Studies in M_2 and M_3 Generations

1. Chlorophyll Mutations

The frequency of chlorophyll mutation is useful in assessing the potency of mutagen and it is also an indicator of factor mutation. It is considered as a dependable index for evaluating genetic effects of mutagenic treatments (Gustafson, 1951 and Monti 1968). The origin of chlorophyll deficiencies is mainly due to mutations in genes, which are responsible for synthesis of photosynthetic pigment. The chlorophyll mutants are usually lethal but semilethal and viable types are also known (Kothekat et.al, 1994).

Apart from the low yield potential, chlorophyll mutants are potentially useful in understanding different physiological functions, pathological invasions and different biochemical reactions (Miller, 1968).

In okra different types of chlorophyll mutants such as *xantha*, *chlorina* and *viridis* could be noticed at the seedling stage in the M_2 generation. It was found that all the three mutagens succeeded in inducing the chlorophyll mutants.

The frequency of chlorophyll mutants in both the varieties of okra could be noticed in a differential manner in all the three mutagens used. Among the three mutagens, the EMS treatment proved to be very much successful in inducing the highest frequency of chlorophyll mutants in variety Parbhani Kranti of okra, while in variety Arka Anamika the EMS again demonstrated that feature.

Similar results have been obtained by D'amato *et al.* (1962), Mackey (1967) in diploid wheat, Heringa (1964) and Hussein *et al.* in pea, Favret (1963) in barley, Sree Ramula (1970) in *Sorghum*, Hussein and Abdella (1974) in *Vicia-faba*, Kothekar (1978) in *Solanum nigrum*, Hakande (1992) in winged bean and Satpute (1994) in safflower. Panda (1973) reported the gamma rays to be more effective in producing the chlorophyll mutations as compared to chemical mutagens in two varieties of wheat.

The chlorophyll mutations werer more effective in gamma ray treatment than EMS in the studies of Jambhale and Nerkar (1982), in okra. Nandpuri *et al.* (1971), Jehangir and Chandrasekharan (1978) have also reported similar results.

Malani *et al.* (1993) reported that the frequency of chlorophyll mutations was highest at 20kR gamma ray treatment in variety selection 2-2 of okra. The chlorophyll mutation frequency revealed an increasing feature with an increase in the dose of mutagen in chilli. (Ramana Rao *et al.*, 1991).

In the present study it has been shown that the *chlorina* mutants were predominant in variety Arka Anamika while the *viridis* mutants demonstrated that situation in variety Parbhani Kranti and there was total absence of *albina* mutants. Similar results were reported by Jehangir and Chandrasekharan (1978) in okra. Kawai (1962) and Oeone (1963) reported high frequency of *albina* in rice.

According to Gustafsson (1963) the ionizing radiations produce high frequency of *albina* mutants whereas chemical mutagens result in high frequency of other types of chlorophyll mutants. Gaul (1964) reported that X-rays and EMS treatments succeeded in inducing more of *viridis* mutants than *albina*. Ehrenberg (1960) recorded *viridis* as the most common type after chemical mutagenic treatments. He further added that there may be a difference in the action of different chemicals on individual chlorophyll gene.

Chlorophyll development seems to be controlled by many genes located on different chromosomes (Swaminathan 1965, Blixt and Gottschalk 1975). A large number of genes are involved in chlorophyll synthesis. Nilan *et al.* (1964) thought that about 250-300 loci are involved in chlorophyll synthesis in barley. According to Guastafsson (1963) 125-250 loci may be required for *albina* and 120 for *viridis*.

Most mutations affecting these loci are inherited as monogenic recessive (Moh and Smith 1951; Moh and Nilan 1956 and Pereau Leroy 1968). Swaminathan (1963) suggested that the genes controlling chlorophyll development are located near the centromere and proximal segments of the chromosomes.

Mutagenic effectiveness and efficiency :-

The mutagenic effectiveness is indicative of the frequency of mutations induced by a unit dose of mutagen. The mutagenic efficiency on the other hand provides an idea of the mutation frequency in relation to biological damage such as lethality, injury, sterility, chromosomal aberrations etc produced through mutagenic treatments (Konzak *et al.* 1965).

Many investigators have made attempts for determining the most effective/ efficient mutagens and the pertinent treatment conditions that would allow the induction of desirable characters in various plants. Ehrenberg (1960) and Kawai (1969) illustrated that the highest mutation frequency induced should be taken into account while comparing mutagenic efficiency.

In the present investigation it was observed that the mutagen EMS proved more effective than SA and gamma ray treatment in both the varieties of okra. It was further observed that the SA proved more effective than gamma rays in regard to different biological parameters.

The highest value of mutagenic efficiency could be observed at lower doses of gamma rays than at higher doses in okra. (Jehangir and Chandrasekharan, 1978 and Malani *et al.* 1993).

In the present study the order of efficiency of mutagens varied with different biological parameters studied. For lethality, the efficiency increased in order SA, gamma rays and EMS in variety Parbhani Kranti, while in variety Arka Anamika the sequence was gamma rays, SA and EMS. For pollen sterility the order was gamma rays, SA and EMS in both the varieties of okra *viz.* Parbhani Kranti and Arka Anamika.

Many workers (Blixt, 1964; Heringa, 1979; Hakande 1992; More 1992; Prasad 1972; Sharma and Sharma 1979; Satpute 1994 and Khandewal 1996) have observed the alkylating agents to be more efficient than gamma rays.

Spence (1965) exhibited in barley that the chemical mutagen SA induced a higher percentage of chlorophyll mutation over gamma rays. Reddy *et al.* (1988) in rice could record a higher efficiency of sodium azide than the gamma rays and the combination treatment. Hakande (1992) noticed NEU as more efficient in regard to lethality and gamma rays to be maximally efficient as far as pollen sterility and mitotic aberrations are concerned in two varieties of winged bean.

2. Viable Mutations

Quite a good number of viable mutants have been observed in M_2 generation of both the varieties of okra. The frequency of viable mutants revealed an increasing trend with gradual increase in dose/concentration of all the mutagens in variety Parbhani Kranti. In variety Arka Anamika the frequency of viable mutants increased with an increasing dose/concentration of EMS and SA treatments while it fluctuated in gamma ray treatment of variety Arka Anamika.

The chemical mutagens EMS and SA showed increasing trend with increasing concentrations in both varieties of okra *viz.* Parbhani Kranti and Arka Anamika.

There are several reports as regards the induction of viable mutants after physical and chemical mutagenic treatments (Down and Anderson, 1956; Matsuo and Onazawa, 1961a, 1961b; Kawai *et al.*, 1961; Patil 1966; Gregory 1968; Blixt 1972; Joshua *et al.*, 1972; Thakare *et al.*, 1973; Chaudhari 1974; Deshpande 1980 Dnyansagar and Kothekar 1983; Datta and Laxmi 1992; Kothekar and Kothekar 1992; More 1992; Satpute 1994 and Khandelwal 1996).

Ehrenberg and Gustafsson (1957) indicated that the frequency of viable mutations and the mutagenic concentration have some definite relationship. Sree Ramulu (1970) inferred that the frequency of mutation is directly related to the concentration of mutagen.

Changes in phenotype is a result of several steps during the course of mutation. Several mechanisms have been proposed to account for the mutability differences. No evidence seems to explain the full range of spectrum realized.

Genter and Brown (1942), Lamprecht (1958), Gunckel and Sparrow (1962), Dnyansagar and Kothekar (1983) indicated that although the genetic material of cells is sensitive to radiation damage, both the primary and secondary physiological effects may be responsible for many morphological changes.

In the present investigation number of early flowering mutants have been observed in M_2 generation of okra. The feature like early flowering has been given paramount importance while planning the breeding strategies in almost all the crop plants. The earliness gets attained mainly due to rapid growth taking place during early stages of ontogeny and initiation of first inflorescence at a very low position on the stem.

Nandpuri *et al.* (1970) reported early flowering mutant in okra. Priadcencu (1962) reported earliness in French bean, Jana (1962) in black gram, Carpenter (1958) in clover, Wellensiek 1959) in Pea, Gaul *et al.* (1966) and Sharma (1979) in barley, and Kothekar and Kothekar (1992) in moth bean reported early flowering mutants. Gottschalk and Wolff (1977) opined that the early flowering mutants could be very much useful for gene ecological studies.

In okra several extreme dwarfs could be observed in the M_2 population. Gaul (1993) reported that cells of germinating shoot and root had mitotic aberrations which can cause inhibitory effect on growth. It was also indicated that destruction of auxins in growth parts of plants resulted in reduced growth (Smith and Kersten, 1942 and Rajput and Quershi 1973).

The dwarf, bushy mutant was reported by Nandpuri *et al.* (1970) in okra and Stubbe (1957) in tomato, Malani *et al.* (1993) in okra, Jehangir and Chandrasekharan (1978) in okra, Priadcencu (1962) in bean, Down and Anderson (1956) in *Phaseolus vulgaris*, Kundu (1980) in Black gram, Deshpande (1980) in *Momordica charantia*, Hakande (1992) in winged bean, More (1992) in *Medicago sativa*, Kothekar (1992) in Moth bean and Satpute (1994) in safflower.

The tall mutants in the present investigation induced by different mutagens showed an appreciable increase in height. Nandpuri *et al.* (1970). Jehangir and Chandrasekharan (1978) and Malani *et al.* (1993) reported the tall mutants in okra through the effect of gamma irradiation.

Similar types of mutants were reported in varied other plants *e.g.* Jehangirdar (1975) in *Foeniculum vulgare*, Khanolkar (1977) in *Carum copticum*, Dnyansager and Kothekar (1983) in *Solanum nigrum* and Satpute (1994) in safflower. Jana (1963) concluded that the tall and twining mutants show a pleiotropic effect and the elongation is possibly a single major event due to which the internodes and leaves become longer. Webber and Gottschalk (1973) observed that the longitudinal increase in cell length was accompanied by an increased number of cells per unit area.

In the present studies a good number of early and late maturing mutants have been distinctly seen at M_2 level. The traits like early flowering and maturity have been always considered important aims in the breeding of almost all crop plants.

Some mutants showing high yielding character were also very much prevalent in the M_2 population of okra. Similar reports have been made in the progeny of gamma ray treated okra by Nandpuri *et al.* (1970). The yield of fruits per plant was higher only in gamma radiation treatment than the chemical mutagen treatment. (Jehangir and Chandrasekharan, 1978 in okra and Gupta *et al.*, 1969 in peas). Tickoo and Jain (1979) with EMS and gamma rays obtained high yielding and disease resistant mutants in *Vigna radiata*, Bhatnagar *et al.* (1979) in *Cicer arietinum*, Pawar *et al.* (1986) in mungbean, Singh and Agrawal (1986) in cluster bean reported high yielding mutants in M_2 generation.

The yield character in plants is genetically controlled. The expression of yield controlling gene is also influenced by environment and many of the high yielding mutants require different conditions than the one by their parental counterpart for better performance. Khan and Hashim (1979) got increased number of branches, pods per plant and yield after using gamma rays, EMS and Hydrazine in green gram.

In the present investigation different mutagens induced long pod mutants in both the varieties of okra *viz.* Parbhani Kranti and Arka Anamika. An appreciable frequency of long pod mutants has been induced by lower and middle doses of DES and gamma rays in different plants (Jehangir and Chandrasekharan, 1978). Similar results were reported by Malani *et al.* (1993) in different varieties of okra.

It has been observed that both the varieties of okra showed marginal differences in their response towards the mutagens pertaining to the frequency of viable mutants.A differential response of different varieties towards mutagenic treatment has been recorded by Marki and Maiyo (1970) in Flax, Korotkova (1971) in wheat, Dnyansagar and Kothekar (1986) in *Momordica charantia*, Hakande (1992) in winged bean and Satpute (1994) in safflower.

The differences in radiosensitivity between varieties and species have been attributed to moisture content of seeds (Caldecott, 1955 and Emery *et al.*, 1970) and oxygen pressure (Ehrenberg *et al.*, 1953). Wallace (1965) correlated chromosomal damage and gene mutation to specific moisture content, which enhanced mutation production at specific locus in oat.

Among the mutants discussed above the dwarf early maturing, long pod and high yielding mutants seem to be the most promising ones. Such mutants can be exploited from the commercial point of view.

Quantitative Characters in M_2 and M_3 Generations

Mutation breeding has become an alternative to conventional breeding since the last three decades with the sole objective of developing better cultivars of economically important crops/vegetables. An increase in the mutation rate by mutagenic agents enhances the chance of getting desired variations and to help accelerate the breeding process as compared to the usual practice of developing novelties.

Almost all economically important characters with which the plant breeder has to deal are polygenic in nature and the genotype for these traits cannot be characterized directly being highly modified by environment. These polygenic characters are governed by a large number of independent genes and are also called metric traits as they can be studied only through appropriate biometrical procedure. Micro mutations may be useful in plant breeding as they might occur much more frequently and the change in the physiological features could be less drastic in them.

The ultimate improvement of any system largely depends on the quantum of genetic variability available within it. The mutagenesis helps to enhance natural mutational rate and to enlarge the genetic variability thereby generating added scope for making further selections. The induction of micromutations in polygenic system controlling the quantitative characters are important for crop improvement.

The major objective in employing induced mutations has been to increase variability within shortest possible period and to develop some such genotypes which could carry attributes of immense economic value.

Quite a good number of researchers have studied the various aspects of quantitative genetics. Some of the pioneers in this field are Gustafsson (1974) and Gaul (1967) in barley, Gregory (1955 and 1956) in peanut, Rawlings *et al.* (1958) in soyabean, Blixt (1961) in pea, Grifith and Johnston (1962) in oat, Sakai and Suzuki (1964) in rice, Brock (1967) in *Arabidopsis*, Goud (1967), Scossirolli (1956 and 1968) and Borojevic and Borojevic (1968) in wheat and Aastveit (1967) in rye. Several investigators have studied critically the induced polygenic variability *e.g.* Bhatia and Swaminathan (1962) in wheat, Daly (1960) in *Arabidopsis*, Chaturvedi and Singh (1978) in Mungbean, More (1992) in *Medicago sativa*, Satpute (1994) in safflower and Khandelwal (1997) in winged bean. All such workers have noted that although the variance changed the mean remained either reduced, enhanced or equal to that of control in all the treated populations. It was noted by them that the variability was different in different generations under different conditions.

Brock (1965) and Gregory (1968) suggested that when the selection is applied, the mean shift in the subsequent generation is much reduced and the variability is greater than the earlier one. According to Bateman (1959) and Brock (1965), the polygenic mutations always follow a particular direction opposite to the previous history of selection.

It has been indicated that the number of factors like pollen sterility, genetic and physiological imbalances along with environmental variations either singly or combination could be the decisive factors in shifting the mean value of yield parameter in negative direction.

In the present studies depression of mean values in M_2/M_3 generations of okra accompanied by an increased variance have been observed in majority of treatments. Gaul and Aastveit (1966) have opined that an increase in the number of variants associated with reduced vitality could be the reasons for some such expressions. Parallel type of results were also obtained by Rawlings *et al.* (1958) in soybean, Swaminathan (1963) in wheat, Sakai and Suzuki (1964) in rice and Gregory (1965) in peanut.

The negative mean shift has been reported in M_2 population of green gram (Shakoor *et al.*, 1978 and Tickoo and Jain, 1979). Rajput (1974) and Hakande (1992) on the other hand have observed a positive as well as negative mean shift for various quantitative characters in M_2/M_3 generations. A few workers like Dahiya (1973) and Chaturvedi and Singh (1978) have, however reported a positive mean shift.

The mean values of different parameters need not necessarily change despite increased variance. This particular feature gets attained on account of induction of plus and minus mutations in equal proportions affecting the quantitative characters (Oka *et al.*, 1958, Kawai *et al.*, 1961 and Matsuo and Onazawa, 1961a, and 1961b). Gaul (1965) and Aastveit (1966) reported that the induced polygenic mutations need not follow any particular direction and they can occur at random.

In the present investigation some of the parameters exhibited a shift in mean values in positive direction in M_2 and M_3 generations of both the varieties of okra. The significant mean shift in positive direction however was evident for parameters in variety Parbhani Kranti like days to maturity in gamma rays (10kR, 20kR) EMS (0.10 per cent and 0.15 per cent) plant height at gamma rays and EMS treatment; first flower appearing node in EMS (0.10 per cent and 0.15 per cent) and SA treatment; pod length, and pods per plant in (0.10 per cent and 0.15 per cent) EMS, green pod yield in all treatments. In variety Arka Anamika, the positive shift in mean could be observed in parameters like: days to maturity in 0.15 per cent EMS plant height in SA and EMS treatments nodes on main stem (EMS, 0.15 per cent), pod length in 5kR gamma ray treatment.

It has been decisively demonstrated by several researches like Gregory (1955), Rawlings *et al.* (1958) and Oka *et al.* (1958) that in progeny population derived from mutagenised parental plants, the variability as regards quantitative characters increases while the mean of characters may or may not change. This increase in variability has been ascribed to mutations of minor genes controlling characters.

In the present study, the variance in M_2 and M_3 generations increased in majority of treatments.The shift in variance was found to be significant in respect of most of the characters in all the mutagenic treatments of both the varieties of okra.The induced variability was quite evident for most of the characters in M_2 and M_3 generations.

Different workers like Dahiya (1973), Mujeeb and Grieg (1973), Rajput (1974), Shakoor *et al.* (1978), Tickoo and Jain (1979); Khan and Hashim (1979), Hakande (1992), More (1992), Satpute (1994) and Khandelwal (1996) have recorded an increased variability in the quantitatively inherited characters in the mutagenised population. Ota *et al.* (1962) opined that there exists a direct relationship between mutagen dose and variance. A wider variability in the yield parameters of mutagen treated French bean has been reported by Swarup and Gill (1968). Similar features in M2 and M3 generations have been recorded by Hakande (1992) in winged bean, More(1992) in *Medicago sativa,* Satpute (1994) in safflower and Khandelwal (1996) in winged bean.

Frey (1965) has shown that the mutagen induced variability for quantitative characters in crop plants is heritable and response to selection is good. He further indicated that the selection in such situations must be carried out with great care as the variability depends mostly on the magnitude of phenotypic expressions produced by the mutations in polygenic loci. In case of okra the enhanced induced variance was quite evident for most of the quantitative characters but the magnitude thereof tended to differ in different mutagenic treatments. From this trend it could be inferred that the mutagens gamma rays; EMS and SA used in present study although acted differentially have definitely proved successful in broadening the genetic base to an appreciable extent. This can very much create additional opportunities for making effective selection and for incorporating the useful variability in conventional breeding programme of okra system for developing its desirable recombinant types.

Chapter 6

Summary and Conclusion of Induced Mutation

The okra botanically described as *Abelmoschus esculentus* (L.) Moench, is a member of family Malvaceae. It is one of the most important fruit vegetables grown throughout the tropics and warmer parts of temperate zone. It has been reported to have nutritive value higher than tomato, egg plant and most cucurbits.

Due to short duration and its photoinsensitive nature and the ability to grow throughout the year, the okra fruit vegetable has been making rapid strides in the human diet of our country.

The okra vegetable fulfills majority of our nutritive requirements besides providing a number of medicinal effects mainly due to the high iodine content in its fruits. For the preparation of medicine the green fruit is used for controlling human diseases like renal colic, leucorrhoea, dysentry and general weakness. The area and production of okra in India is quite satisfactory as compared with other countries.

In view of this background the utility of novel approaches has been thought quite crucial for enhancing the yield potential of okra in our country. By keeping in mind the excellent utility potential of induced mutations in plant improvement programmes, it was thought very much appropriate and relevant to initiate the mutation breeding work in okra for achieving its genetic improvement.

Looking at the immense economic importance of the okra crop as a fine source of green vegetable with good medicinal value, the improvement in its yielding potential has been considered necessary. This crop though unstable in its area and productivity in our state is gradually gaining importance chiefly due to its multifaceted utility value.

The survey of related literature indicated that very little attention has been paid by the scientists of our country in regard to creation of new variability through the established method of induced mutation in okra.

In the present study two varieties of okra *viz.* Parbhani Kranti and Arka Anamika were used to induce genetic variability. For this study a physical mutagen(gamma rays) and two chemical mutagens(EMS and SA) were tried.

Studies pertaining to mutation breeding of okra were spread over three generations.

Studies in M_1 Generation

The different parameters studied in M_1 generation were 1) seed germination, 2) survival of plants at maturity, 3) leaf morphological changes, 4)chlorophyll deficient sectors in leaves and 5) pollen sterility.

1. Seed Germination

All the three mutagens had an inhibitory effect on seed germination in variety Parbhani Kranti whereas in case of Arka Anamika the germination increased in SA treatment only and revealed a decrease in gamma ray and EMS treatments, respectively.

2. Survival of Plants at Maturity

The survival of plants after gamma ray treatment demonstrated inhibitory effect in Parbhani Kranti and Arka Anamika. It indicated increased values in majority of the EMS and SA treatments in both the varieties of okra.

3. Leaf Morphological Changes

It was observed that in all the treatments the leaves of plants exhibited variations comprising:

a) Suppression of leaf lamina and
b) Small leaf size.

The frequency of plants carrying leaf morphological changes varied in different mutagenic treatments tried in the present study.

4. Chlorophyll Deficient Sectors

The chlorophyll deficient sectors were of different types such as yellow (*xantha*), white (*albina*), lightgreen (*viridis*) and yellow green (*chlorina*). Such sectors were located at the margins of leaves or spread through lamina producing beautiful chimeric appearances. They were noticeable at higher frequency in the material treated with SA 0.01 per cent and 0.03 per cent, while the material treated with EMS and gamma rays indicated relatively lower frequency values in two varieties of okra.

5. Pollen Sterility

The percentage of pollen sterility was maximum in the gamma ray (5kR) treatment in Parbhani Kranti and Arka Anamika varieties. The EMS and SA induced increasing pollen sterility in Parbhani Kranti and Arka Anamika with the gradual rise in mutagenic concentration.

Studies in M_2 Generation

In the M_2 generation which was raised as progeny lines, a wide spectrum of chlorophyll mutants like *xantha, chlorina* and *viridis* could be detected. Their frequency showed wide fluctuations with dose/concentration of different mutagens. The highest frequency was induced by gamma rays (10kR) in variety Parbhani Kranti, while in variety Arka Anamika it was EMS (0.10 per cent) which attained that feature.

From this data and the data on biological damage in M_1 generation, the relative effectiveness and efficiency of the three mutagens used was assessed. It was SA (0.01 per cent) in variety Parbhani Kranti and SA (0.02 per cent) in Arka Anamika, which could be noted as maximally effective concentrations.

The order of mutagenic efficiency varied with different biological parameters studied. In variety Parbhani Kranti the efficiency of lethality revealed the order EMS > gamma rays > SA, while in Arka Anamika the efficiency of lethality demonstrated the order EMS > SA > gamma rays.

For pollen sterility the order was EMS > SA > gamma rays in variety Parbhani Kranti, whereas in variety Arka Anamika the efficiency of pollen sterility revealed the order EMS > SA > gamma rays.

The mutations rates were calculated taking the mean values of efficiency for each treatment. This particular parameter revealed the variable order for the mutagens as the mutagens have different values for lethality and pollen sterility.

The M_2 population revealed quite a good number of viable/morphological mutants having desirable characteristics. The spectrum of such mutants was found to be broad in case of both the varieties of okra.

The different viable mutants obtained were of the types: dwarf and tall, long pod, early and late maturing, high yielding and branching types, in both the varieties of okra. The frequency of such viable mutants demonstrated a variable trend at different doses/concentrations of majority of the mutagens.

Most of the viable mutants could be observed as breeding true in subsequent M_3 generation. Some of these mutants can be very well utilized on commercial basis in view of the positive attributes possessed by them. In this regard the dwarf early maturing, long pod and high yielding mutants of okra would be of immense value. A good amount of scope exists for exploiting these mutants by incorporating them in the breeding programme to help develop desirable recombinants.

The data on eleven quantitative characters such as 1) Days to flowering 2) Days to maturity 3) Plant height 4) Nodes on main stem 5) Green pod yield (in gm) 6) First flower appearing node 7) Pod length (cm) 8) 100 seed weight (in gm) 9) Pods per plant 10) Seeds per pod and 11) First picking duration were collected in M_2 and M_3 generations of okra.

The statistical analysis of pertinent data was carried out to understand the effect of mutagens in shifting the mean and variance in either direction. The mean variance and coefficient of variation were computed. Most of the mutagenic treatments in both the varieties of okra succeeded in showing a significant positive shift in mean values.

The analysis of water soluble seed protein content (per cent) was undertaken in some of the promising M_3 mutants lines of okra by Lowrys method. The degree of fluctuation induced in protein level of various mutant was quite marked in variety Parbhani Kranti and Arka Anamika. This feature yet again supported the relative differential genotypic status of the two varieties of okra used in present study.

From the above it is concluded that the various mutagenic treatments used in the present study have very much succeeded in inducing genetic variability with significant alteration in growth and metabolism of the plant body. The results obtained decisively demonstrate the usefulness and effective potential of induced mutational approach in genetic improvement of okra for recovering superior mutant plant types having enhanced green pod yield besides protein content.

Chapter 7

Biochemical Investigation of Viable Mutants

Introduction

In modern plant breeding one of the major trends has been to support the traditional studies by biochemical investigations with the hope of obtaining a better estimate of the breeding value of a variety strain.

The nutritional improvement of food crops through breeding has attracted lot of attention in last two decades, (Milner, 1972). The importance of this aspect can be realized from the fact that the IAEA in collaboration with the FAO initiated a broad programme on this topic since 1968.The objective was to increase the worlds protein resources by means of selection and mutations. Jain (1969) and Swaminathan and Jain (1972) have indicated that the existing genetic variability in different crop plants can be utilized effectively for isolating the nutritionally improved strains.

It has been agreed by different researchers that the ultimate importance of cereals, pulses, oil crops, crucifers and several other crops is simply not restricted to the number and weight of seeds produced. But there are some specific substances stored in the seeds (*viz.* Proteins, oils, fats, starch, sugar etc.) which are of considerable importance. In particular, the seed proteins and seed oils are of immense application from human/animal nutrition point of view.

The approach of mutation breeding has been found to be quite fruitful as regard the qualitative/quantitative improvement of some such materials that are crucial from nutritional angle. Quite a good number of investigators like Sakai and Sazuki (1964), Borojevic (1965), Gaul (1967), Chaudhari (1974), Gaikwad (1975), Deshpande (1980) Kothekar (1983), Hakande (1992) and More (1992) have made useful contributions for enhancing the content of vital substances like protein, vitamin-c, oil and fat by resorting to mutational approach in different plants.

The okra being a vegetable crop of immense economic value, studies were planned to look into the biochemical aspects of different promising M3 mutant lines of the system.

Materials and Methods

Extraction of Seed Proteins

Mature seeds were washed with water, dried and ground to make a fine powder. The mature seed powder was defatted with hexane, air dried and stored at 4degree c. Seed proteins from mature defatted seed powder were kept for extraction in 1.6 proportion of distilled water with 1 per cent (PVP) polyvinyl polypyrrolidone. The suspension was centrifuged at 12000, rpm at 4 °C for 20 minutes to remove the particulate matter and clear supernatant was used for further protein value was expressed as mg/gm (1951) of defatted seed powder.

Results (Tables 74-75)

Soluble seed protein content in control was 126 mg/gm in variety Parbhani Kranti. Variability was revealed by mutants in the value of extractable seed protein content. Five out of Eight mutants exhibited increase in soluble protein content. The highest value was 131 mg/gm found in early flowering mutant while the lowest value was 124 mg/gm in tall mutant type in variety Parbhani Kranti.

In Arka Anamika variety soluble seed protein content in control was 122 mg/gm.The values of soluble seed protein content increased in five mutants out of eight mutants. The highest value was 127 mg/gm found in high yielding mutant type while in dwarf mutant.

Macromutants showed variability in the soluble seed protein content in both the varieties of okra. It ranged between 124 mg/gm to 131 mg/gm in variety Parbhani Kranti and 114 mg/gm to 127 mg/gm in variety Arka Anamika.

Table 74: Water Soluble Seed Protein Content in M_3 Mutants of *Albelmoschus esculentus* (L.). Moench.

Variety: Parbhani Kranti

Sl.No.	*Name of Mutant*	*Extractable Protein Content (mg/gm)*
1.	Control	126
2.	Dwarf	127
3.	Tall	124
4.	Branched	130
5.	Early flowering	131
6.	Long pod	125
7.	Early maturing	129
8.	Late maturing	126
9.	High yielding	130

Table 75: Water Soluble Seed Protein Content in M_3 Mutants of *Albelmoschus esculentus* (L.). Moench.

Variety: Arka Anamika

Sl.No.	*Name of Mutant*	*Extractable Protein Content (mg/gm)*
1.	Control	122
2.	Dwarf	114
3.	Tall	120
4.	Branched	126
5.	Early flowering	120
6.	Long pod	121
7.	Early maturing	123
8.	Late maturing	125
9.	High yielding	127

Discussion

The seed protein content in okra has revealed an increase in some of its M_3 mutants induced through the gamma rays EMS and SA treatments. As against this, some mutants in variety Parbhani Kranti and Arka Anamika demonstrated a declining nature as well.

Several investigators have made attemps for altering the protein content in induced macro and micro-mutants of different crop plants (Haq *et al.*,.1970; Gaul *et al.*, 1973; Chaudhari 1974; Walther *et al.*, 1975; Hakande 1992 and More 1992).

Different reports are available regarding the effect of radiation and chemical mutagens on the protein content (Swaminathan *et al.*, 1968; Tanaka 1969;Siddiq *et al.*, 1970; Rabson and Bhatia 1975 and Reddy 1977). Srivastava and Roy (1981) noticed a decreasing trend for crude protein content after gamma irradiation in *Solanum melongena*.

Gottschalk and Muller (1970) opined that the improvement in protein content and composition by way of genetic approaches is a permanent improvement. It has been very much appreciated in recent years that the protein synthesis in seeds as regulated by a series of specific genes and the changes in proteins are produced due to genetic alteration in most of events. It is agreed that the total seed protein content, the protein pattern and the relative proportion of different protein fractions are influenced by different mutants genes.

Variability in protein after treatment with reference to its qualitative/ quantitative aspects was reported by different researchers like Varghese and Swaminathan (1966) in wheat, Tanaka and Takagi (1970) in rice, Gottschalk and Muller (1970) in pea, Sjodin (1971) *in Vicia*, Chaudhari (1974) in *Trigonella foenum-graecum*, Thakare *et al.* (1981) in green gram, Bhargava and Khalotkar (1983) in barley and Kothekar *et al.* (1994) in winged bean.

Investigators like Bhagwat *et al.* (1979) have observed existence of a negative correlation between increase in protein content and some other traits like days to maturity, number of grains and their size and the grain yield.

An enhancement in protein content after gamma ray treatment has been recorded by Palmarcuk (1961) in red clover.

Gottschalk and Muller (1978) have recorded an improvement in protein content and its composition through genetic manipulations. The pertinent observations have conclusively indicated little scope for a simultaneous increase in protein content and yield.

Bibliography

Aastveit, K. (1966): Use of induced barley mutants in a cross-breeding programme. In: "Mutations in plant breeding (Proc. Panel, Vienna, 1996) IAFA, Vienna: 7-14.

Aastveit, K. (1967): Induced mutations affecting quantitative characters in autotetraploid in rye populations. *Rad. Bot.*, 7: 363-368.

Aman R. D. (1968): A Model of seed dormancy. *Bot Rev.*, 34: 1-31.

Anon (1988): Personal communucation from agricultural and processed food products Export development Authority (PAEDA), Bhikaji cama.

Anonymous (1980): Fertilizers and their applications to crops in nigera fertilizers use series No. 1 191; Lagos, Nigeria, federal ministry of agricultural.

Anonymous (1990): Pakage of practice for vegetables and fruit crops. Panjab Agri. Uni. Ludhiana. (in Press).

Anwar S. V., Tejovathi G., Khadeer M. A., Seeta P. and Rajendra Prasad B. (1993): Tissue culture and mutational studies in safflower (*Carthamus tinctorius* L.) Proceedings, Third International Safflower Conference, Beijing, China, 124-136.

Apparao S. and Jana M. K. (1976): Leaf Mutation induced in black gram by X-rays and EMS. *Environ. Exp. Bot,* 16:151154.

Arora S. K. (1980): Genetic analysis for some quantitative and quanlitative characters in okra. Ph. D. dissertation. Himachal Pradesh Krishi Vishvidyalaya Solan.

Arumugam R., Chelliah S. and Muthukrishnan C. R. (1975): A. manihot. A source of resistance to bhindi yellow vein mosaic *Madras Agric. J.* 62: 310-312.

Arumugam R. and Muthukrishnan C. R. (1977): CO-1 Bhendi A variety for Tamil Nadu. *Madras Agric. J.* 64: 811-812.

Arumugam R. and Muthukrishnan C. R. (1979): Gene effects of some quantitative characters in okra. *Indian J. Agric. Sci.* 49: 602-604.

Ashri A. and Herzog Z. (1972): Different physiological sensitivity of peanut varieties to seed treatment with DES and EMS. *Rad. Bot.* 12 : 173.

Auerbach C. (1943): Drosophila melanogaster New mutants chemically induced mutants and rearrangements. *Dros. Int. Serv.* 17: 48-50.

Auerbach C. (1947): Proceedings of the Royal Society Edin. 62:307-320.

Basha S. K. and Rao P. G. (1988): Gamma ray and Sodium azide induced heterophylly of bhindi. *J. Nuclear Agric. Biol,* 17: 133-136.

Basha S. K. and Rao P. G. (1989): Leaf size induced by X-ray irradiation and associated changes in protein and chlorophyll in bhindi. *J. Nuclear Agric. Biot,* 18: 187-189.

Bateman A. J. (1959): Induction of polygenic mutation in rice. *Int. J. Rad. Biol.* 1: 425-427.

Bhagwat S. G., Bhatia C. R., Gopalkrishna T., Joshua D. C., Mitra R. K., Narhari Bhargava Y.R. and Khalatkar A. S. ('1983): Induced mutations for seed protein improvement in *Hordeum vulgare* L. *Curr. Sci.* 52(10):493-494.

Bhatia C. R. and Swaminathan M. S. (1962): Induced Polygenic variability in bread wheat and its bearing on selection procedure Z. Pflanzenzuecht. 43: 317-326.

Biradar A.B. 2004. Gamma Ray and Ethyl Methane Sulphonate (EMS) Induced Mutation Studies in Pigeonpea [*Cajanus Cajan* (L.) Millsp.]. MPKV Rahuri. Ph.D Thesis. India.

Blixt S. (1961): Quantitative studies of induced mutation in peas, V, Chlorophyll mutations. *Agric, Hort. Genet,* 19: 402-447.

Blixt S. (1964): Studies on induced mutations in peas VIII Ethylene imine and gamma ray treatment of the variety witham wonder. *Agric. Hort. Genet.* 22 : 171-183.

Blixt S. (1967): Studies on induced mutations in peas XXIV genetically conditioned differences in radiation sensitivity-2, *Hereditas* 59: 303-326.

Blixt S. (1972): Mutation Genetics in *Pisum, Agric,. Hort. Genet,* 21: 178-216.

Bora K.G. and Patil S.H. and Subbaiah K.C.(1961): X-ray and neutron induced meiotic irregulation in plants with special reference to Arachis *hypogea and Plantago ovata*. In Proc. Symp. On "The effects of Ionising Radiation on Seeds". IAEA, Vienna: 203-216.

Borojevic K. (1965): The effect of irradiation on the number of kernels per spike in wheat. In: "Report FAO/IAEA Tech meeting rome, 1964, pergamon press, oxford: 505-513.

Borojevic K.and Borojevic S. (1968): Response of different genotypes of *Triticum aestivum* to mutagenic treatments. In : "Mutations in plant breeding" II, IAEA, Vienna:15-46.

Borojevic K. and Van Herten A. M. (1978): Applications of mutation breeding methods in the improvement of vegatatively programmed crop. Elsevier, Amsterdam: 316.

Brock R.D. (1965a): Induced mutation affecting quantitative characters. In: "The use of Induced mutation in plant breeding ". *Rad. Bot. Suppl.,* 5:451-464.

Brock R.D. (1965a): Induced mutation affecting quantitative characters. In soybean. Crop Sci. 1: 187-190.

Brock R.D. (1965b): Response of *Trifolium subteranum* to X-rays and thermal neutron. *Rad. Bot.* 5:543-555.

Brock R. D. (1967): Quantitative variations in *Arabidopsis thaliana* induced by ionizing radiations. *Rad. Bot.* 7: 193-203.

Broertjes G. And Van Hartan A. M. (1978): Application of mutation breeding methods in the improvement of vegetatively propagated crops. Elsevier, Amsterdam: 316

Caldecott R. S. (1955): Effects of hydration on X-ray sensitivity in *Hordenum Rod. Res.,* 3: 316-330.

Candolle A. P. DE (1824): Hibiscus. In: Prodromous systematics naturalis regril vegetables, 1: 446-455.

Candolle A. P. DE (1883): Origine des plantes cultivers. Paris: 150-151.

Caradona C., Daurte R. and Mansini S. (1960): The use of artificial radiation in agriculture and its application to kindney bean in Colombia. Agriculture trop. 16: 514-523.

Carpenter J. A. (1958): The induction of mutation in subterranean clover by X-irradation, *J. Australian Inst. Agric. Sci.* 24: 39-44.

Chaturvedi S. N. and Singh V. P. (1978): Induced mutagenic effects of EMS in mung bean (*Phaseolus aureus* Roxb.) by DMSO. *J. Cyto. Genet.* 13: 116-119.

Chaturvedi V. and Pant M.C. 1988. Effect of feeding fenugreek (*Trigonella foenum-graecum*) leaves on focal excretion of total lipids and sterols in the normal albino rabbits. *Curr. Sci.* 57:81-82.

Chaudhari U S. (1974): Cytogenetic studies in *Trigonella foenum-graecum* Linn. Ph. D. Thesis, University of Bombay.

Chevalier A. (1940): L'origine la culture et les usages de cino Hibiscus de la section Abelmoschus. *Rev. Bot. Appl.,* 20: 319-328, 402-419.

Chopra V. L. and Sharma R. P. (1985): In genetic manipulation for crop improvement, Oxford.

Dahiya B. S. (1973): Improvement of mung bean through induced mutations. *Ind. J. Genet.,* 33: 460-468.

Daly K. (1960): The induction of quantitative variability induced by gamma radiation in *Arabidopsis thaliana. Genetics,* 45: 983.

Das M.L., Kabir S.M.R., Begum S. and Ahmed Z.U. 2004. Radio-sensitivity and selection of elite mutants of mungbean [*Vigna radiata* (L.) Wilczek]. Bangladesh *J. Bot.* 33(1): 41-45.

Datar V. V. and Astapure J. U. (1984): Reaction of brinjal variety, F1-hybrids and solanum species to little leaf diseas. *J. Maharashtra Agric. Uni.* 9(3):325-334.

Datta S. K. and Laxmi V. (1992): Induced morphological mutations in Fenugreek. *J. Indian Bot. Soc.* 71: 65-68.

Datta S. K. (1994): Radiation induced ornamental varieties developed at NBRI, Lucknow. Proc. DAE-BRNS Symp. On "Nuclear Application in Agriculture, Animal Husbandry and Food Preservation. March 16-18, 1994, New Delhi:11-12

Davis D. R. and Evans H. J. (1966): The role of genetic damage in radiation induced cell lethality. *Adv.Rad.Biol.*2: 243-353.

Deshpande N.M. (1980): The effect of gamma rays and chemical mutagens a in Momordica charantia L. Ph. D. Thesis, University of Nagpur.

Dhapke and Meshram (1981): Quoted in: Advances in Horticalture Vol-5, vegetable crop part-1.

Dhillan T. S. and Sharma B. R. (1979): Interspecific hybridisation in okra (*Abelmoschus sp.*). *Genet. Agrar* 36:247-256.

Dnyansagar V. R. and Kothekar V. S. (1983): Gamma rays induced morphological mutants in dipoid *Solanum nigrun* L. Proc seminar Current Approaches in Cytogenetics, Patna, 189-195.

Down E. E. and Anderson A. L. (1956): Agronomic use of X-ray induced mutants. *Science,* 124: 223-224.

Dutta O. P. (1969): Breeding orka for quality yield and resistance to virus and insct pests. Half yearly report. Inst. Hort. Res. Banglore.

Dutta O. P. (1984): Breeding orka for quality yield and resistance to virus and enation leaf curl virus. Annu. Rep. IIHR, Banglore. 43.

Ehrenberg L., Lundquist U. and Nybom N. (1953): Irradiation effects, seed soaking and oxygen pressure in barley. *Hereditas* 39: 493-503.

Ehrenberg L.and Gustafsson A. (1957): On the mutagenic action of ethylene oxide and diepoxybutane in barley. *Hereditas* 43: 595-602.

Ehrenberg L. (1960): Induced mutation in plants: Mechanism and principles. *Genet. Agric.* 12: 364-389.

Emery D. A., Boardman E. G. and Stucker R. E. (1970): Some observations on the radiosensitivity of certain varietal and hybrid genotypes of cultivated peanuts. *Rad. Bot.* 10: 267-272.

Evans H.J. and Sparrow A.H. 1961. Nuclear factors affecting Radiosensitivity, II. Dependence on nuclear and chromosome structure and organization; Brookhaven Symp, *In Biol.* 14:101-127.

Favert E. A. (1963): Genetic effects of single and combined treatments of ionizing radiations and EMS on barley seeds. Proc. Intern. Barley Genetics Symp. Washington 1963: 68-81.

Fishbein L., Flamm W. G. and Falk H. L. (1970): Chemical mutagenesis. Academic Press New York.

Frey K. J. (1965): Mutation breeding for quantitative attributes In: "The use of Induced Mutations in Plant Breeding", (Rep. FAO/IAEA Tech. Meet. Rome, 1964): 465-475.

Gadwal V. R., Joshi A. B. and Iyer R. D. (1968): Interseptic hybrids in Ablemoschus through ovule and embryo culture. Indian J. Genet. Plant Breeding 28: 269-274.

Gager C. S. and Blackeslee A. E. (1927): Chromosome and gene mutations in Datura following exposure to radium rays. Proc. *Natl Acad. Sci. U. S.* 13:75-79.

Gaul H. (1957): De wirkung von Rontgemstrahein in virbindung mit, CO_2, colchicin and hitreauf. Genee Z. *Pflgucht* 38: 397.

Gaul H. (1958a): Uber dio gegenseitige, Unabhangig keit der chromosomes and punkl mutationen z. *Pflanzeneucht*, 40: 151-188.

Gaul H. (1958b): Present aspects of induced mutations in plant breeding. Euphytica, 7:275-278.

Gaul H. (1964): Mutations in plant breeding for forage and grain. *Rad. Bot.* 4(3): 151-232.

Gaul H. (1965): The concept of macro and micro mutations and results on induced mutations in plant breeding (FAO/IAEA meeting Rome) Pergamon Press: 407.

Gaul H. and Aastveit K. (1966): Induced variability of culm length in different genotypes of hexaploid wheat following X-irradiation and EMS treatment. Comptemp. *Agric.*, 11-12: 263-276.

Gaul H., Ulonska E. and Sato M. (1966): EMS induced genetic variability in barley: "The problem of EMS-induced sterility, a method to increase the efficiency of EMS treatment", In: mutations in plant breeding" IAEA, Vienna, pp.63-64.

Gaul H. (1967): Studies on populations of micromutations in barley and wheat without and with selection. In : "Induced mutation and their utilization. Proc. Symp. Erwin Bauer Gedachtnisvorlesungen Gatersleben, 1966: 269-281.

Gaul H., Frimmel G., Gichner T. and Ulonska E. (1972): Efficiency of mutagenesis in: "Induced mutations and Plant Improvement". (Proc. Meet. Buenos Aires 1970) IAEA, Vienna: 121-139.

Gaul H. (1977): Sterility in: "Manual on Mutation Breeding" Tech. Rep. Series No. 119, FAO/IAEA, Vienna 1977: 96-98.

Gelin O. E. A. (1954): X-ray mutants in peas and vetches. Acta Agr. Scand. 4: 558-568.

Gelin O., Ehrenberg L. and Blixt S. (1958): Genetically conditioned influence on radiation sensitivity in peas. *Agric. Hort. Genet.* 16: 78-102.

Genter C. F. and Brown H. M. (1942): X-ray studies on the field bean. *J. Heredity*, 32: 39-44.

Girifith D. J. and Johnstone T. C. (1962): The use of irradiation technique in oat breeding. *Rad. Bot.*2: 41-52.

Gladstone J. S. and Hunter A. W. S. (1959): Effect of seed moisture storage on the growth and survival of X_1 tomato plants. *Can. J. Genet. Cyt.* 1: 339-46.

Gottschalk W. and Muller H. (1978): Monogenic alteration of seed protein content and Protein Pattern in X-ray induced Pisum mutants from" Improving plant protein by nuclear Techniques", IAEA, Vienna:201-215.

Gottschalk W. and Wolff G. (1983): The behavior of protein rich pisum mutant in crossing experiments. In: Gottschalk W., Muller H. P. (Eds.) "Seed Protein, Biochemistry, Genetics, Nutritive value", Nishoff, The Hague: 403-425.

Goodspeed T. H (1929): The effect of X-rays and radium on species of the genus Nicotiana. *J. Heredity*, 20: 243-259.

Goud J. V. (1967): Induced mutations in bread wheat. *Indian J. Genet.* 27: 40-55.

Gregory W. C. (1955): X-ray breeding of peanut (*Arachis hypogea* L.), *Agron. J.* 47: 396-399.

Gregory W. C. (1956): Induction of useful mutations in the peanut. In: "Genetics in Plant breeding". Proc. Brookhavean Symp. *Biol.* 9: 177-190.

Gregory W. C. (1965): Mutation frequency, magnitude of change and the probability of improvement in adoption. In: "The use of induced Mutations in plant Breeding" (FAO/IAEA Tech. Meet. Rome, 1964): 429-441.

Gregory W. C. (1968): A radiation breeding experiment with peanuts. *Rad. Bot.*8: 81-147.

Grewal (1971): Advances in Horticalture vol. 5, Vegetable Crop. Part 1.

Grewal B. S., Nandpuri K. S. and Kumar J. C. (1973): Effect of sowing dates spacing and picking of green pods on yield and quality of okra seed. Punjab Hortic. J. 13: 247-52.

Grubben G.J. H. (1977): Okra (in) Tropical vegetables and their genetic resources. IBPGR, Rome: 111-114.

Gunckel J. E. and Sparrow A. H., Marrow J. B. and Christensen E. (1953): Vegetative and floral morphology of irradiated and non-irradiated plants of Tradescantia paludosa. Amer. J. Bot., 40: 317-332.

Gupta C. R. and Yadhav R. D. S. (1984): Genetic variability and path analysis in chilli. *Genetic Agraria*, 38:425-432.

Gupta M. N. (1979): Use of gamma irradiation in the production of new varieties of perennial portulaca. In: Proc. Symp. On radiations and Radiomimetic substances in mutation Breeding",: 206-214.

Gupta P. K. and Yashvir (1975): Mutagenic effects of individual and combined treatment of gamma rays and EMS in okra (*Abelmoscus esculentus* (L.) Moench.) *J. Cytol. Genet.* 9 and 10: 93-97.

Gustafsson A. (1938): Studies in the genetic basis of chlorophyll formation and the mechanism of induced mutations, Lund. Uni. Arsskr. N. F. Adv. 5.

Gustafsson A. (1940): The mutation system of the chlorophyll apparatus. Lund. Uni. Asrak. N. P. *Adv.* 36: 1-40.

Gustafsson A. (1947): Mutations in agricultural plants. *Heredits* 33.: 1-100.

Gustafsson A. (1951): Induction of changes in genes and chromosome II mutations, environment and evolution. Cold spring. Harbour symp. *Qyabt. Bio.* 16: 263-281.

Gustafsson A. (1963): Productive mutations induced in barley by ionizing radiation and chemical mutagens. *Heredits.* 50: 211-263.

Gustafsson A. (1969): A study on mutations induced in plants. In: "Induced mutation in plant, IAEA, Vienna: 9-31.

Hakande T. P. (1992): Cytogenetical studies in *Psophocarpous tetragonolobous* (L.) DC Ph. D. Thesis, Marathwada University, Aurangabad, India.

Harsulkar A.M. 1994. Studies on the mutagenic effects of pesticides in barley; Ph.D. Thesis, BAM University Aurangabad, India.

Haq M. S., Ali S. M., Maniruzuman A. F. M. Mansur A. and Islam R (1970): Breeding for earliness, high yield and disease resiatance in rice by means of induced mutations from "Rice Breeding with induced mutations" IAEA, Vienna: 77-78.

Heringa R. T. (1965): Mutation research in pea. *Euphytica,* 13: 330-336.

Hugo de Vries (1901): Die mutations theoric 1. Von viet and Co., Leipzig: 1-648

Hussein H. A. S. and Abdaulla M. M. F. (1974): Effects of single and combined treatment of gamma rays and EMS on the M1 fertility and M2 chlorophyll mutations in *Vicia faba* l. *Egypt J. Genet. Cytol.* 3(2): 246-258.

Ignacimuthu S. and Babu C.R. 1989. Induced variation in protein quality in the wild and cultivated urd and mungbean. *Ind. J. Gent.* 49(2): 171-181.

Ignacimuthu S. and Babu C. R. 1992. Induced variation in pod and seed traits of wild and cultivated beans. *J. Nucl. Agric. Biol.* 21(4): 286-292.

Jahagirdar H. A. (1975): Cytogenetic studies *in Foeniculum vulgare* Mill. Ph. D. Thesis, University of Nagpur.

Jahagirdar K. S. ans Chandrasekaran P. (1978): Comparative mutagenic effects of gamma rays and diethyl sulphate in bhindi. *Madras Agri. J.* 65(4):211-217.

Jain H. K. (1969): World Sci. News 6(3): 30-32.

Jambhale N. D. and Nerkar Y. S. (1982): Induced amphidiploidy in the cross A. esculentus (L.) Moench x *A. manihot* (L.) medik spp. Manihot *Genet. Agrar,* 36:19-22.

Jana. M. K. (1962): X-ray induced mutations of *Phaseolus mungo* L. II. Sterility and Vital mutants. *Genet. Iber.* 14: 71-104.

Jana. M. K. (1963): X-ray induced mutations of *Phaseolus mungo* L. II. Chlorophyll mutation. *Caryologia,* 16: 685-692.

Jha. A. and Mishra J. N. (1955): Yellow vein mosaix or bhindi (*Hibiscus esculentus* L.) Proc. Bihar Acad. *Agri. Sci.* 4: 129-130.

Johnson Edna L. (1928): Growth and germination of sunflower as induced by X-rays. *Am. J. Bot.* 15.

Joshi A. B., Gadwal V. R. and Hardas M. W. (1974): Okra. In: Hutchinson J. B. (ed.) Evolutionary studies in world crops. Diversity and change in Indian subcontinent. Camb. Uni. Press. London:99-105.

Joshi A. B. and Hardas M. W. (1956): Alloploid nature of okra *Abelmoschus esculentus* (L.) Moench. Nature, 178: 119.

Joshi A. B., Singh H. B., Joshi B. S. and Hardas M. L. (1956): Alloploid nature of okra *Abelmoschus esculentus* (L.) *Moench. Nature,* 178: 119.

Joshi A. B. and Hardas M. W. (1976): Okra. In: Simmonds N. W. (ed.) Evolution of crop plants. Longman's London. 194-195.

Joshua D. C., Rao C. and Gottschalk W. (1972): Evolution of leaf shape in jute. *Ind. J. Genet.* 32: 392-399.

Kale P. B. and Gonge V. S. (1993): Effectivity of gamma rays induced mutations in okra. *PKV. Res. J.* Vol. 17(1).

Kaul M. L. H. and Bhan A. K. (1971): Effect of mutagens on rice seedlings. In Inter. Symp. "Use of Isotopes and Radiations in Agriculture and Animal Husbandry Research", New Delhi.

Kawai T., Sato H. and Masima J. (1961): Short culm mutations in rice induced by P32. Proc. Symp. "Effect of Ionizing Radiations on Seeds", IAEA, Vienna: 565-573.

Kawai T. (1969): Relative effectiveness of physical and chemical mutagens "Induced mutation in plants". Proc. Series, IAEA, Vienna: 137-152.

Khan I. A. and Hashim M. (1979): A comparative account of mutagenesis with gamma rays, EMS and HZ involving quantitative parameters in green gram. In: The role of induced mutations in Crop Improvement. Prco. Symp. DAE, India: 420-424.

Khan S. and Wani M.R. 2005. Comparison on the effect of chemical mutagens on mungbean. *Adv. Plant Sci.* 18(II): 533-535.

Khan A.A., Ansari M.Y.K., Bhat T.A., Shahab D. and Wani N.A. 2006b. Screening of mutants in M3 generation induced by chemical mutagens in broad bean (*Vicia faba* L.). *Biosci. Biotech. Res. Asia.* 3(1a): 245-248.

Khandelwal A. R. (1996): Cytological Biochemical and Mutational Studies in *Psophocarpus tertragonolobus* L. (DC.). Ph.D Thesis, Dr. B. A. Marathwada University, Aurangabad. MS. India.

Khanolkar S. M. (1977): Cytogenetic studies in *Carum copticum* Ph. D. Thesis, University of Nagpur.

Kharkwal M.C. 1999. Induced mutations in chickpea (*Cicer arietinum* L.) III. Frequency and spectrum of viable mutations. *Indian J. Genet. Plant Breed.* 59(4): 451-464.

Kivi E. I. (1962): On sterility and other injuries in *Melandrium* irradiation with X-ray and gamma rays. *Ann. Acad. Sci. Enn. Ser.* A, IV, 56: 1-56.

Koli N.R. and Ramkrishna K. 2002. Frequency and spectrum of induced mutations and mutagenic effectiveness and efficiency in fenugreek (*Trigonella foenumgraecum* L.). *Indian J. Genet. Plant Breed.* 62(4): 365-366.

Konzak C. F. (1956): Induction of mutation for disease resistance in cereals, Brookhaven Symp. *Biol.*, 9: 157.

Konzak C. F., Nilan R. A., Wagner J and Foster R. J. (1965): Efficient chemical mutagenesis. In: "The use of induced mutation in plant breeding", *Rad. Bot.* (Supl.), 5: 49-70.

Kolhe A. K. and D'curz R. C.F. (1966): Inheritance of pigmentation in okra. *Ind. J. Genet. Plant. Breed.*, 23: 112-117.

Konzak C.F. (1957): Genetic effect of radiation on higher plants, *Rev. Biol.* 32: 27-45.

Korotkova A. P. (1971): Role of genotype in induced mutagenesis of winter wheat S-KH. *Biol.* 6(6): 831-835.

Kothekar A. V. and Kothekar V. S. (1992): Promising mutants in moth bean. *Marathwada University J. Sci.* 19: 1-2.

Kothekar V. S. (1978): Mutational studies in *Solanum nigrum* L. Ph.D. Thesis, University of Nagpur.

Kothekar V. S. and Dnyansagar V. R. (1985): Polyploidy and sensitiviy to radiations in *Solanum nigrum* L. *J. Cytol. Genet.* 20: 79-88.

Kothekar V. S. (1987): Differential mutagenic sensitivity in *Coriandrum sativum* Linn. *Curr. Sci.* 56: 491-492.

Kothekar V. S.(1989): Differential radiation sensitivity in moth bean. *Curr. Sci.* 58(13): 758-760.

Krishnaswami R. (1967): Induced viruescent bud mutation in *Gossypium hirsutum Crop Sci.* 36: 387.

Kulkarni R. S. and Thimmappiah (1977): Inheriatance of number of pods in bhindi. *Agric. Res. J. Kerala* 15:174-175.

Kumar S. and Natarajan (1970): The response of bhindi to irradiation, Tamil Nadu Agri. Uni. Coimbatore.

Kumar A., Mishra M.N., and Kharkwal M.C. 2007. Induced mutagenesis in blackgram (*Vigna mungo* L. Hepper). *Indian J. Genet.* 67(1): 67-50.

Kundu S. K. (1980): EMS and Gamma ray induced variability in black gram (*Vigna mungo*) M. Sc. Thesis.

Kuwada H. (1967): Mutations induced in Abelmoscus manihot by 60CO gamma rays and X-rays. Ikushugaku zashi/*Jap. J. Breed.* 17: 205-211.

Lamprecht, H. (1958): Uber grundlegende gene fur die Gestaltung Loherer pflanzen sowie under reue and bekannte,Rontgen Muyanten. *Agric. Hort. Ganet. Landskrona*, 57: 317-324.

Larik, A. S. (1975): Induced polygenic mutations in wheat (*Triticum aestivum* L.). *Genetica Polonica* 16:153-160.

Lawley, P. D. (1973): Reaction of NMU with 32 p-labeled DNA, evidence for formation of phosphotriesters. *Chem. Biol. Interaction,* 7:127-130.

Lawley, P. D. (1974): Some chemical aspects of dose response relationship in alkylating mutagenesis. *Mut. Res.,* 23: 283-295.

Lindstrom E. W. (1933): Radium induced variations in the tomoto. *J. Hered.* 24: 128-137.

Loveless A. (1966): Genetic and allied effects of alkylating agents. Butterworths, London: 270.

Lowry O. H., Rosebrough N. J. Farr A. L. and randall R. J. (1951): protein measurement with the folin phenol reaquent. *J. Biol. Chem,* 193: 265-275.

Mackey J (1954): Mutation breeding in polypoid cereals. *Acta Agric Scand.* 4: 549-557.

Mackey J (1956): induced mutation. *Brookhavan Symp. Biol.* 9: 141-152.

Mackey J (1961): Mutation and plant breeding, NAS-NRC, 891:336-364.

Mackey J. (1967): Physical and chemical mutagenesis in relation to ploidy lavel. Abhandl. Deutsch. Akad. Wiss. *Berlin KL of Med.* 1967(2): 185-197.

Mahajan R and Sharma B. R. (1979): Resistance to root knot nematodes (Meloidogyne incognita) in okra. *Veg. Sci.,* 6: 65-66.

Malani S. S., Kale P. B. and Gonge V. S. (1993): Effectivity of gamma rays induced mutations in okra *PKV Res. J. Vol.* 17(1).

Malik Y. S. (1968): Genetic variability and correlation studies in okra M.Sc. Thesis. Haryana Agric. Uni. Hissar.

Malik Y. S., Singh S. S. and Pandita M. L. (1981): Effect of salinity on okra (*Abelmoschus esculentus* (L.) Moench) *Haryana J. Horticulture Sci.,* 10: 136-241.

Mamidwar R. B., Nerkar Y. S. and Jambhale N. D. (1979): Cytogenetics of interspecific hybrids in the genus Abelmoschus. *Indian J. Hered.* 11: 35-39.

Manohar and Solanki (1980): Advances in Horticalture vol-5 vevetable crop part-1.

Manjaya J.G. and Nandanwar R.S. 2007. Genetic improvement of soybean variety JS 80-21 through induced mutations. *Plant Mut. Rep.* 1(3): 36-40.

Manapure Prema and Patil Shanti. 1997. Induced mutants in blackgram. *J. Soils and Crops.* 7(2): 208.209.

Mitra P.K. and Bhowmik G. 1999. Studies in the frequency and segregation of induced chlorophyll mutations in *Nigella sativa* L. *Adv. Plant Sci.* 12: 125-129.

Mundhe B.F. 2008. Gamma rays and ethyl methane sulphonate (EMS) induced mutation studies in soybean [*Glycine max* (L.) Merill.], Ph.D.Thesis, University of Pune. India.

Marki A. and Maiyya Morea Blano (1970): Indatsirovanie mututsi U'lana (*Linum usitatissimum* L.) gamma izochenlem I etilmutoen sul fanatom. *Genetika,* 6(1): 24-28.

Masters M. T. (1875): In: Hooker J. D. (ed.) Flora of British India Ashford Kent, 1: 320-348.

Matsuo T. and Onazawa Y. (1961a): Studies on mutation induced by radiations and chemicals –2. Comparison of frequency of mutation induced by x-rays, neutrons and diepoxy buetan. *Jap. J. Breed,* 11: 295-299.

Matsuo T. and Onazawa Y. (1961b): Mutation induced in rice by ionizing radiations on seeds. IAEA. Vienna: 495-501.

Meherchandani M. (1975): Effect of gamma radiation on the dormant seeds of *Avena Rad. Bot.,* 15: 439-445.

Mehta Y. R. (1959): Vegetable growing in Uttar Pradesh. *Bure. of Agric. Inf.* Lucknow.

Miller F. R. (1968): Inheritance of white stripe and rolled leaf characteristics in *Sorghum biocolor* L. Moench. *Crop Sci,* 8: 769-777.

Milner M. (1972): Nutritional improvement of food legumes by breeding. Proc. of a Symp. Sponsored by protein Advisory group of United Nations. New York.

Moh C. C. (1971): Mutation breeding in seed coat color of bean (*Phaseolus vulgaris* L.) *Euphytica,* 20: 119-125.

Monti L. M. (1968): Mutation in peas induced by diethyl sulfate and X-rays, *Mut. Res.* 5: 187-191.

More A. D. (1992): Cytogenetical studies in *Medicago sativa* L. Ph. D. Thesis, Marathwada University, Aurangabad, India.

Morris R. (1952): Transmitted pollen and chromosomal aberrations induced in maize by tassel exposure in a nuclear reactor. *Amer. J. Bot.* 39: 452-457.

Mujeeb K. A. and Greig J. K. (1973): Gamma irradiation induced variability in *Phaseolus vulgaris* L. *Rad. Bot. 13*: 121-126.

Muller H. J. (1927): Artifical transmutation of the gene. Science, 66:84-87.

Nadarajan N., Ramalingam R.S. and Sivasamy N. 1985. Biological effect of mutagenic treatments in *Cajanus cajan* (L.) Millsp. *Madras agric. J.* 72(6): 301- 305.

Nadkarni K. M. (1927): Indian Materi Medica Nadkarni and Co. Bombay.

Nath P., Velaryudhan S. and Singh D. P. (1987): Vegetables for the Tropical Region, ICAR, New Delhi.

Nandpuri K. S., Sandhu K. S. and Randhava K. S. (1971): Effect of irradiation on variability of okra (*Abelmoschus esculentus* (L.) Moench) J. Res. P.A. U. 8: 183-188.

Narahari P. and Bhatia C. R. (1975): Breeding for seed protein improvement by using Molecular techniques. IAEA, Vienna: 23.

Narahari P., Bhatia C. R., Gopalkrishna T. and Mitra R. K. (1976): Mutation induction of protein variability in wheat and rice. Evaluation of seed protein alteration by mutation breeding, IAEA, Vienna: 119-127.

Narayana K.P. and Konzak C. F. (1969): Influence of chemical post treatments on the mutagenic efficiency of alkylating agents. In: Induced Mutation in Plants IAEA, Vienna: 28.

Newcombe H. B. (1971): Adv. Genet., 16:240-290.

Nilan R. A. (1956): Factors governing plants radio-sensitivity. *Proc. Atomic Energy comm.* Report TID-7512: 151-162.

Nybom N. (1961): The use of induced mutations for improvement of vegetatively propogated plants. In: Mutation and Plant Breeding" (Symp.) *Natl. Acad. Sci. Natl. Res. Council Pub.* 891: 252-294.

Oehlkers N. (1943): Indiklina Abstammunge-U Verbangslehne, 81: 311-134.

Oka H. I, Hayashi J. and Shiofiri I. (1958): Induced mutation of polygenes for quantitative characters in rice. *J. Hered.* 49: 11-14.

Ota T., Sugiyama K. and Itatavia I. (1962): Studies on variation of quantitative characters by gamma radiations. *Japan J. Breed.,* 11: 244-260.

Palamarcuk A. S. (1961): The effect of ionizing radiation and giberelline on red clover and the effect after of the radiation on the F_1 and F_2. *Ukrain J. Bot.,* 18: 49-61.

Panda B.S. (1973): Relative efficiency of single and combination mutagenic treatments on various dwarfing genotypes of bread wheat. Ph. D. Thesis, IARI, New Delhi.

Patil S. H (1966): Mutation induced in groundnut by X-rays. *Ind. J. Genet. And Plant breed.,* 26A:334-348.

Pele S.K. and Howard A. 1955. Effect of various doses of X-rays on the number of cells synthesizing DNA. *Rad Res.* 3: 135-142.

Prasad M.V.R. (1974): A comparison of the mutagenic effectiveness and efficiency of gamma rays, EMS and NG. *Ind. J. Genet.,* 32: 360-367.

Rabson R. and Bhatia C. R. (1975): In Proc. II Int. Winter wheat Conf. Zagreb. Yugoslavia: 441.

Raddy C.S and Husan M.V. and Jagdish C.A. (1988): Role of plant tissue and cell culture in crop improvement. In: Recent Advances in Genetics and Cytogenetics", Eds. Irfan A. Khan and S.A. Farook: 431-446.

Raghuvanshi S. S. and Singh D. N. (1979): Comparative sensitivity of different varieties of *Impatiens balsamina* L. *Cytologia,* 44:111-121.

Rajan S. (1940): *Proc. Ind. Acad. Sci.,* B12: 62.

Rajput M.A. (1974): Increased variability in M2 of gamma irradiated mung bean. *Rad. Bot.,* 14: 85-89.

Ramanna M.S. and Natarajan A.T. (1965): Studies on the relative mutagenic efficiency of alkylating agents under different conditions of treatments. *Ind. J. genet.* 25: 24-25.

Ramanna Rao, Y. V. Rao, M.V.B. Ravindra Babu, V. Rama kumar, P. V. and Subramanyam D. (1991): Efficiency and Effectiveness of physical and chemical mutations in M2 generation of chilli. *Nuclear Agric. Biol.* 20(4): 255-259.

Randhava G. E. and Purewal S. S. (1947): Studies in Hibiscus esculentus. Chromosome and pollination study. *Indian J. Agric. Sci.* 17: 129-136.

Rao J. L (1983): Studies on mutagenesis in okra (*Abelmoscus esculentus* L. Moench) Ph.D. Thesis A.P. Agril. University.

Rao M. V. (1991): Inaugural Address. Sesond Int. safflower conference, Hyderabad:7-12.

Rao and Bidari (1976): A source of resistance to bhindi yellow vein mosaic. Quoted by Sharma and Arora in. Advances in Horticulture vol-5. Vegetable crop: part-1. Advances in Horticulture vol-5. Vegetable crop: part-1.

Rao S. and Sathyvati (1971): Influence of environment on combining ability and genetic components in bhindi, Gentetic Polonica, 18:141-147.

Rao T. S. and Ramu P. M. (1977): A study of correlation and regression coefficient in Bhindi, Current Research No. 6: 135-137.

Rao Rama Mohan D., Reddy T.V. and Seetharami V. 1983. Induction of seedling injury in two varieties of rice by gamma rays, sodium azide alone and in combination. *Indian J. Bot.* 7(1): 64-67.

Reddy V.R.K. and Gupta P.K. 1989. Biological effects of gamma rays and EMS in Hexaploid *Triticale. Acta Bota. Indica.* 17: 76-81.

Rapoport I. A. (1946): Carbonly compounds and the chemical mechanism of mutation O. R. (Doklady) *Acad. Sci.,* USSR, 54: 65-67.

Rawlings J.O., Hanway D.G. and Gardner G.O. (1958): Variation in quantitative characters of soybean after seed irradiation. *Agron. J.* 40: 424-528.

Reddy G.M. (1977): EMS induced grain shape mutations in some local cultivars of rice. II *Riso* 26: 187-192.

Ross W. C. J. (1962): Biological alkylating agents;. Butterworths, London.

Roy K and Jana M.K. (1972): EMS induced early muturation in rice. *Curr. Sci.* 42(3): 101-102.

Sabnis T. S. and Mehta T. R. (1981): *Proc. Natl. Acad. Sci.* India, 12: 76.

Sagade A.B. 2008. Genetic improvement of urdbean through mutation breeding, Ph. D. Thesis, University of Pune. Pune; India.

Sakai K. and Suzuki A (1964): Induced mutation and pleiotropy of genes responsible for quantitative characters in rice. *Rad. Bot.,* 4: 141-151.

Sandhu (1974): A source of resistance to bhindi yellow vein mosaic, Quoted in: Advances in Horticalture Vol-5, vegetable crop part-1.

Sathyamurthy J. (1983): Study on the effect of fraction and combination of mutagenic treatments in Lablab. *Purpureas* (L.) sweet M.Sc. (Ag.) Thesis, TANU, Coimbatore.

Sato L. and Gaul H. (1967): Effect of ethyl methanesulfonate on fertility in barley, *Rad Bot.* 7: 7-14.

Satpute R. A. (1994): Mutational studies in safflower (*Carthamus tictorious* L.) Ph. D. Thesis, Dr. B. A. Marathwada University, Aurangabad, India.

Sax K. (1941): The behevior of X-ray induced chromosome aberrations in *Allium* root tip cells. *Genetics*, 26: 418-425.

Sax K. and Swanson C. P. (1941): Differential sensity of cells to X-rays. *Am. J. Bot.* 28:52-59.

Scossirolli R. E. (1965): Values of induced mutations for quantitative characters in plant breeding. *Rad. Bot.*, 5 (Suppl): 443-450.

Scossirolli R. E., Palenzona D. L. and Scossirolli Pellegrini (1966): Studies on induction of new genetic variability for quantitative traits by seed irradiation and its use for wheat improvement. Mutations in Plant Breeding. IAEA Vienna: 197-229.

Scossirolli R. E. (1968): Selection experiments in a population of *Triticum durum* irradiated with X-rays. In Mutations in plant breeding IAEA, Vienna: 205-217.

Shakoor A., Ahsan-ul-haq and Siddiq M. (1978): Induced variation in mung bean (*Vigna radiata* L. Wilczek). *Environ. And Exptl. Bot.* 18: 169-175.

Sharma B.R. and Arora S. K. (1993): Advances in Horticulture vol-5. Vegetable crop: part-1.

Sharma S. K. and Singh B. (1979): Mutagenic effectiveness and efficiency of gamma rays, N-nitroso-N-methyl urea in lentil *J. Sci. Ind. Res. 39*: 516-520.

Sharma S.K., Sood R. and Pandey D.P. 2005. Studies on mutagen sensitivity, effectiveness and efficiency in urdbean [*Vigna mungo* (L.) Hepper]. *Indian J. Genet. Plant Breed.* 65(1): 20-22.

Siddiq E. A., Kaul A. K., Puri R. P., Singh V. P. and Swaminathan M. S. (1970): Mutagen induced variability in protein characters in *Oryza sativa*. *Mut. Res.*, 10(1): 81-84.

Sigurbjornsson B. and Micke A. (1969): Progress in mutation breeding. Proc. "Induced mutation in plants", IAEA, Vienna: 673-698.

Singh B. P and Choudhary J. B. (1972): Studies on irradiated gaur. *The Nucleus*, 15(1): 16-22.

Singh D. (1983): Biometrical and genetical study in okra. Ph. D. dissertation, PAU, Ludhiana.

Singh H. B.(1957-58): Intervariatal hybridization resistance to yellow vein mosaic, Plant Introduction Section. Division of Botany. IARI, New Delhi.

Singh H. and Gill S. S. (1988): Effect of time of sowing and spacing on seed yield of okra. *J. Res. P.A. U.*, Ludhiana, 25(1):44-48.

Singh K., Malik Y. S., Kalloo and Mehrotra N. (1974): Genetic variability and correlation studies in bhindi. *Veg. Sci.* 1: 47-54.

Singh M. and Chaturvedi S. N. (1987): Effectiveness and efficiency of mutagens alone or in combination with dimethyl sulfoxide in *Lathyrus. sativus* L. *Ind. J. Agri. Sci.* 57(7): 503-507.

Singh and Sikka (1955): New variety for West Bengal. Plant introduction section, IARI, New Delhi.

Sirsov V. A. and Sain S. S. (1973): The variability of leguminous crops under the influence of gamma irradiation. Trans. *Moscow Soc. Nat. Biol. Sect.* 23:159-163.

Sjodin J.J. (1962): Some observations in X-1 and X-2 of *Vicia faba* L. after treatment with different mutagens. *Hereditas,* 48: 565-586.

Sjodin J.J. (1971): Induced morphological variation in *vicia faba* L. *Hereditas* 67: 155-180.

Smith G. F. and Kerstein H. (1942): Auxin and callines in seedling from X-rayed seeds. *Amer. J. Bot.* 29: 785-792.

Solanki I.S. and Sharma B. 1994. Mutagenic effectiveness and efficiency of gamma rays, ethylene imine and N-nitroso N-ethyl urea in macrosperma lentil (*Lens culinaris* Medik). *Indian J. Genet. Plant Breed.* 54: 72-76.

Sonavane A.S. (2000): Genetic improvement of winged bean through mutation breeding Ph. D. Thesis Dr. Babasaheb Ambedkar Marathawada University, Aurangabad.

Sparrow A. H. (1961): Types of ionizing radiations and their cytogenetic effects. Mutations and plant breeding. NAS-NRC, 891: 55-119.

Sparrow A. H. and Konzak C. F. (1958): Camellia Calcutta: 425-452.

Sparrow A. H. and Woodwell G. M. (1962): Predication of the sensitivity of plants to chronic gamma irradiation. *Rad. Bot.* 2: 9-26.

Spence R. K. (1965): The influence of sodium azide on the biological effects of ionizing radiation in moist barely seeds. M. Sc. Thesis Washington State Univ. Pullman, Wash.

Sree Ramalu K. (1970): Mutation in Sorghum. *Mut. Res.* 10: 198-205.

Sree Ramalu K. (1971): Chemical mutagenesis in *Sorghum* Proc. *Ind. Acad. Sci. B.* 174: 161-173.

Srivastava D. P. and Roy S. K. (1981): Effect og gamma radiation on nitrogen protein content in different parts of egg plant cultivars. J. Indian Bot. Soc. 60: 252-256.

Stadler L. G. (1928a): Mutations in barley induced by X-rays and radium. *Science,* 68:186-187.

Stadler L. G. (1928b): National Academy of Science, USA 14: 69-75.

Stadler L. G. (1930): Some genetic effects of X-rays in plants. *J. Heredity,* 21: 3-19.

Steucart J. (1960): Unterchungen uber die wirkung von rontgenstran-len auf Rispenhirse (pancum milliaceum Z.) mach einmaliger undie mehrfacher Bestrahlung 1. Die Mutabillitat nach einmalgan Bestrahulng Sawie die Beziehungn Swischen Ausbesebeginn und selecktion serologe. Z. *Pflanzenzuecht,* 43: 85-105.

Sudhakaran I. V. (1971): Meiotic abnormalities induced by gamma rays in *Vinca rosea*. Linn. *Cytologia*. 36: 67-79.

Sun L. and Singer B. (1975): The specificity of different classes of ethylating agents towards various sites of *HeLa* Cell DNA in *vitro* and in *vivo*. *Biochemistry*, 14(8): 1795-1802.

Swaminathan M.S., Chopra V.L. and Bhaskaran S. 1962. Chromosomes aberrations, frequency and spectrum of mutations induced by EMS in barley. *Indian J. Genet. Plant Breed.* 22: 192-207.

Swaminathan M.S. (1963): Evaluation of the use of induced micro and macro mutations in the breeding of polyploid plants. In: "Use of induced mutations in plant breeding", FAO/IAEA, Tech. *Meet.*: 619-641.

Swaminathan M.S. (1964): A comparison of mutation induction in diploids and polyploids. FAO/IAEA Technical Meeting on "Use of induced mutations in plant breeding", pp. 619-641.

Swaminathan M.S., Siddiq,E.A. Sarin U.N. and Varughese G. (1968) : Studies on the enhancement of mutation frequency and identification of mutation in plant breeding and phylogenetic significance of mutations in plant breeding in some cereals. Mutation in plant breeding II, IAEA,pp. 223.

Swaminathan M.S. (1969): " Induced mutations in plants" IAEA., Vienna: 719-736.

Swaminathan M.S. (1971): *J. Indian Bot. Sci.* 50A: 416-429.

Swaminathan M.S. and Jain H. K. (1972): Nutritional improvement of food legumes by breeding PAG. UN system:69.

Swarup V. and Gill H. S. (1968): X-ray induced mutations in French bean and roses. Intern. Symp. "Use of Isotopes, radiations in agricalte and Animal Husbandry Research". New Delhi: 13-25.

Tambe A.B. 2009. Induction of genetic variability in soybean [*Glycine max* L.) Merrill.] for yield contributing traits. Ph.D. Thesis. University of Pune. India.

Tanaka S. (1969): Some useful mutations induced by gamma irradiation in rice In: "Induced mutations in plants", (Proc. Symp. Pullman 1969), IAEA, Vienna: 517-527.

Tanaka S. and Takagi Y. (1970): protein content of rice mutants. In "Improving plant protein bt nuclear techniques", IAEA, Vienna,:55-62.

Thakare R. G. and Joshua D. C. and Rao N. S. (1973): Induced viable mutations in *Corchorus olitorius* L. *Ind. J. Genet*. 33: 204-220.

Thakare R. G. and Pawar S. E., Joshua D. C., Mitra R. and Bhatia C. R. (1981): Variation in some physiological components of yield in induced mutants of mung bean. Proc., Int. symp. Vienna 9-13 March, IAEA, 213-226.

Thandapani (1978): Quoted in Advances in Horticulture vol-5. Vegetable crop: part-1.

Tickoo J. L. and Jain H. K. (1979): Breeding high yielding varieties of mung (*Vigna radiata* (L) Wilczek) through mutagenesis Proc. Symp. On " The role of induced mutations in crop Improvement." Dept. of Genetics, Osmania University, Hyderabad. September 10-13, 1989:198-203.

Tunwar and Singh (1985): A new variety or Tamil Nadu, Tamil Nadu Agric. Uni., Coimbatore central seed committee.

Ugale (1976): Cytological study of *Ablemoschus esculentus* (L.) Moenck. Advances in Horticalture Vol-5, vegetable crop part-1.

Varughese G. and Swaminathan M. S. (1966): Changes in protein quantity and auality associated with a mutation for amber color in wheat. *Curr. Sci.* 35:469-470.

Vijayraghwan C. and Wariar U. A. (1946): Evaluation of high yielding Hybrid bhindi. Proc. 33rd Indian Sci. Congress Abstr. Sect. 10 (Abstract 30).

Venkataramani K. S. (1952): Preliminary studies on some intervarietal crosses and hybrid vigour in *Hibiscus esculentus J. Madras Uni.,* 22: 183-200.

Wallace A.T. (1965): Increasing the effectiveness of ionizing radiations in induced mutations at the vital locus controlling resistance to the Fungus *Helminthosporium victorie* in Oats. *Rad. Bot.* 5 (Suppl.): 237-250.

Walther H. Gaul H. Ulonska E. and Seibold K. H. (1975): Variation and selection of protein and lysine mutants in spring barley, from "Breeding for seed protein improvement using nuclear techniques", IAEA,Vienna: 79-89.

Wanjari K. B. and Kutarekar D. R. (1977): Study of M_1 parameters in different treatments of gamma rays and chemical mutagens in Sorghum. *J. Cytol. Genet.*12: 55-61.

Wani M.R. and Khan S. 2005. Studies on induced chlorophyll mutations in mungbean [*Vigna radiata* (L.) Wilczek]. *Indian J. App. Pure Biol.* 20(1): 47-50.

Wellensiek S. J. (1959): Neutronic mutations in pea. *Euphytica* 8: 209-215.

Wokes F. and Organ J. G. (1943): *Biochem J.* 37: 259-265.

Yashvir (1975): Induced quantitative mutation in okra, plant height mutant. Proc. Indian Acad. Sci. 81 B(4): 181-185.

Zacharias M.(1956): Mutations versche an Kultirflanzen.VI. Rontgenbest ranlungen der sosabohne(*Glycine soja* L.) *Zilcchter* 26: 321-328.

Zammore L. (1965): Effect of mutagenic agents in *Vicia sativa*. Comparison between effects of EMS, EI and X-rays on induction of chlorophyll mutation. In: "The use of induced mutations in plant breeding". *Rad. Bot.* (Suppl.) 5: 205-213.

www.ingramcontent.com/pod-product-compliance
Ingram Content Group UK Ltd.
Pitfield, Milton Keynes, MK11 3LW, UK
UKHW021953270726
14060UKWH00002B/492